AF528601

# FISH DISEASES

## Basic Handbook

# FISH DISEASES
## Basic Handbook

**by**

**Sooraj Ravindran**
**Anupama A**

Wordpen Academics
Publishing Science That Matter

**© 2026 PUBLISHER**

*Information contained in this work has been published by Wordpen Academics and has been obtained by its author(s)/editor(s) from sources believed to be reliable and are correct to the best of their knowledge. However, the publisher and its author(s) make no representation of warranties with respect of accuracy or completeness of the contents of this book, and shall in no event be liable for any errors, omissions or damages arising out of use of this information and specifically disclaim any implied warranties or merchantability or fitness for any particular purpose.*

*All rights reserved. Including the right to translate or to reproduce this book or parts thereof except for brief quotations in critical reviews.*

**ISBN: 978-93-6084-299-4 (Hardback)**

**Published by :** **WORDPEN ACADEMICS**
**4736/23, Ansari Road, Darya Ganj,**
**New Delhi – 110 002**
**Phone: +91-011-35004336**
**E-mail: info@wordpenacademics.com**

***Printed at*** **:** **Replika Press Pvt. Ltd.**

*Exclusive Distributor:* ***Bio-Green Books, New Delhi***

# PREFACE

Fish Diseases: Basic Handbook offers a thorough exploration of the common diseases that affect fish species, with an emphasis on the prevention, diagnosis, and management of these conditions in both wild and farmed aquatic environments. As the global demand for fish and seafood continues to grow, the health of fish populations is increasingly important not only for sustainable fisheries but also for aquaculture industries. This book serves as an essential resource for students, researchers, veterinarians, aquaculture professionals, and fish farmers seeking to understand the biology of fish diseases and the strategies required to maintain healthy fish populations.

This handbook covers the wide spectrum of fish diseases, from bacterial, viral, and fungal infections to parasitic infestations, environmental stressors, and genetic disorders. It provides detailed descriptions of each disease, including symptoms, causative agents, transmission pathways, and effective treatment methods. With an emphasis on practical approaches, the book discusses diagnostic techniques, including laboratory tests, field observations, and modern technologies used to identify and monitor fish health.

The growing reliance on aquaculture as a major food source has introduced new challenges related to fish disease management. The book addresses these challenges by exploring disease control strategies in both intensive and extensive aquaculture systems. We discuss biosecurity measures, vaccination programs, antibiotic resistance, and sustainable practices to prevent disease outbreaks that could harm both the industry and the environment.

In addition to disease management, this book highlights the importance of a holistic approach to fish health, considering factors such as water quality, nutrition, genetics, and the impact of climate change on disease dynamics. By integrating these aspects, readers will gain a deeper understanding of how various factors contribute to fish health and how to effectively address disease threats in a balanced and sustainable way.

We extend our gratitude to the researchers, veterinarians, and aquaculture experts whose work has shaped our understanding of fish health and disease management. It is our hope that Fish Diseases: Basic Handbook will serve as a foundational resource for those working to ensure the health and well-being of fish populations, whether for conservation or for food production.

**Sooraj Ravindran**
**Anupama A**

# CONTENTS

# 1

# INTRODUCTION

Fish farming is the principal form of aquaculture, while other methods may fall under mariculture. Fish farming involves raising fish commercially in tanks or enclosures, usually for food. A facility that releases young (juvenile) fish into the wild for recreational fishing or to supplement a species' natural numbers is generally referred to as a fish hatchery. The most common fish species raised by fish farms are salmon, carp, tilapia, European seabass, catfish and cod.

There is an increasing demand for fish and fish protein, which has resulted in widespread overfishing in wild fisheries. Fish farming offers fish marketers another source. However, farming carnivorous fish, such as salmon, does not always reduce pressure on wild fisheries, since carnivorous farmed fish are usually fed fishmeal and fish oil extracted from wild forage fish. In this way, the salmon can consume in weight more wild fish than they weigh themselves. The global returns for fish farming recorded by the FAO in 2008 totalled 33.8 million tonnes worth about $US 60 billion.

## Major categories of fish aquaculture

There are two kinds of aquaculture: extensive aquaculture based on local photosynthetical production and intensive aquaculture, in which the fish are fed with external food supply.

## Extensive aquaculture

Limiting for growth here is the available food supply by natural sources, commonly zooplankton feeding on pelagic algae or benthic animals, such as crustaceans and mollusks. Tilapia species filter feed directly on phytoplankton, which makes higher production possible. The photosynthetic production can be increased by fertilizing the pond water with artificial fertilizer mixtures, such as potash, phosphorus, nitrogen and micro-elements. Because most fish are carnivorous, they occupy a higher place in the trophic chain and therefore only a tiny fraction of primary photosynthetic production (typically 1%) will be converted into harvest-able fish.

A second point of concern is the risk of algal blooms. When temperatures, nutrient supply and available sunlight are optimal for algal growth, algae multiply their biomass at an exponential rate, eventually leading to an exhaustion of available nutrients and a subsequent die-off. The decaying algal biomass will deplete the oxygen in the pond water because it blocks out the sun and pollutes it with organic and inorganic solutes (such as ammonium ions), which can (and frequently do) lead to massive loss of fish.

Another option is to use a wetland system such as that of Veta La Palma.

In order to tap all available food sources in the pond, the aquaculturist will choose fish species which occupy different places in the pond ecosystem, e.g., a filter algae feeder such as tilapia, a benthic feeder such as carp or catfish and a zooplankton feeder (various carps) or submerged weeds feeder such as grass carp.

Despite these limitations significant fish farming industries use these methods. In the Czech Republic thousands of natural and semi-natural ponds are harvested each year for trout and carp. The large ponds around Trebon were built from around 1650 and are still in use.\

## Intensive Aquaculture

In these kinds of systems fish production per unit of surface can be increased at will, as long as sufficient oxygen, fresh water and food are provided. Because of the requirement of sufficient fresh water, a massive water purification system must be integrated in the fish farm. A clever way to achieve this is the combination of hydroponic horticulture and water treatment, see below. The exception to this rule are cages which are placed in a river or sea, which supplements the fish crop with sufficient oxygenated water. Some environmentalists object to this practice.

The cost of inputs per unit of fish weight is higher than in extensive farming, especially because of the high cost of fish feed, which must contain a much higher level of protein (up to 60%) than cattle food and a balanced amino acid composition as well. However, these higher protein level requirements are a consequence of the higher food conversion efficiency (FCR—kg of feed per kg of animal produced) of aquatic animals. Fish like salmon have FCR's in the range of 1.1 kg of feed per kg of salmon whereas chickens are in the 2.5 kg of feed per kg of chicken range. Fish don't have to stand up or keep warm and this eliminates a lot of carbohydrates and fats in the diet, required to provide this energy. This frequently is offset by the lower land costs and the higher productions which can be obtained due to the high level of input control.

Essential here is aeration of the water, as fish need a sufficient oxygen level for growth. This is achieved by bubbling, cascade flow or aqueous oxygen. Catfish, Clarias spp. can breathe atmospheric air and can tolerate much higher levels of pollutants than trout or salmon, which makes aeration and water purification less necessary and makes *Clarias* species especially suited for intensive fish production. In some *Clarias* farms about 10% of the water volume can consist of fish biomass.

The risk of infections by parasites like fish lice, fungi (Saprolegnia spp.), intestinal worms (such as nematodes or trematodes), bacteria (e.g., Yersinia spp., Pseudomonas spp.), and protozoa (such as Dinoflagellates) is similar to animal husbandry, especially at high population densities. However, animal husbandry is a larger and more technologically mature area of human agriculture and better solutions to pathogen problem exist. Intensive aquaculture does have to provide adequate water quality (oxygen, ammonia, nitrite, etc.) levels to minimize stress, which makes the pathogen problem more difficult. This means, intensive aquaculture requires tight monitoring and a high level of expertise of the fish farmer.

Very high intensity recycle aquaculture systems (RAS), where there is control over all the production parameters, are being used for high value species. By recycling the water, very little water is used per unit of production. However, the process does have high capital and operating costs. The higher cost structures mean that RAS is only economical for high value products like broodstock for egg production, fingerlings for net pen aquaculture operations, sturgeon production, research animals and some special niche markets like live fish.

Raising ornamental cold water fish (goldfish or koi), although theoretically much more profitable due to the higher income per weight of fish produced, has never been successfully carried out until very recently. The increased incidences of dangerous viral diseases of koi Carp, together with the high value of the fish has led to initiatives in closed system koi breeding and growing in a number of countries. Today there are a few commercially successful intensive koi growing facilities in the UK, Germany and Israel.

Some producers have adapted their intensive systems in an effort to provide consumers with fish that do not carry dormant forms of viruses and diseases.

## SPECIFIC TYPES OF FISH FARMS

Within intensive and extensive aquaculture methods, there are numerous specific types of fish farms; each has benefits and applications unique to its design.

### Cage System

Fish cages are placed in lakes, bayous, ponds, rivers or oceans to contain and protect fish until they can be harvested. The method is also called "off-shore cultivation" when the cages are placed in the sea. They can be constructed of a wide variety of components. Fish are stocked in cages, artificially fed, and harvested when they reach market size. A few advantages of fish farming with cages are that many types of waters can be used (rivers, lakes, filled quarries, etc.), many types of fish can be raised, and fish farming can co-exist with sport fishing and other water uses. Cage farming of fishes in open seas is also gaining popularity. Concerns of disease, poaching, poor water quality, etc., lead some to believe that in general, pond systems are easier to manage and simpler to start. Also, past occurrences of cage-failures leading to escapes, have raised concern regarding the culture of non-native fish species in open-water cages. Even though the cage-industry has made numerous technological advances in cage construction in recent years, the concern for escapes remains valid.

Recently, copper alloys have become important netting materials in aquaculture. Copper alloys are antimicrobial, that is, they destroy bacteria, viruses, fungi, algae, and other microbes. In the marine environment, the antimicrobial/algaecidal properties of copper alloys prevent biofouling, which can briefly be described as the undesirable accumulation, adhesion, and growth of microorganisms, plants, algae, tubeworms, barnacles, mollusks, and other organisms.

The resistance of organism growth on copper alloy nets also provides a cleaner and healthier environment for farmed fish to grow and thrive.

In addition to its antifouling benefits, copper netting has strong structural and corrosion-resistant properties in marine environments.

Copper-zinc brass alloys are currently (2011) being deployed in commercial-scale aquaculture operations in Asia, South America and the USA (Hawaii). Extensive research, including demonstrations and trials, are currently being implemented on two other copper alloys: copper-nickel and copper-silicon. Each of these alloy types has an inherent ability to reduce biofouling, cage waste, disease, and the need for antibiotics while simultaneously maintaining water circulation and oxygen requirements. Other types of copper alloys are also being considered for research and development in aquaculture operations.

## Irrigation Ditch or Pond Systems

These use irrigation ditches or farm ponds to raise fish. The basic requirement is to have a ditch or pond that retains water, possibly with an above-ground irrigation system (many irrigation systems use buried pipes with headers.) Using this method, one can store one's water allotment in ponds or ditches, usually lined with bentonite clay. In small systems the fish are often fed commercial fish food, and their waste products can help fertilize the fields. In larger ponds, the pond grows water plants and algae as fish food. Some of the most successful ponds grow introduced strains of plants, as well as introduced strains of fish.

Control of water quality is crucial. Fertilizing, clarifying and pH control of the water can increase yields substantially, as long as eutrophication is prevented and oxygen levels stay high. Yields can be low if the fish grow ill from electrolyte stress.

### *Composite Fish Culture*

The Composite fish culture system is a technology developed in India by the Indian Council of Agricultural Research in the 1970s. In this system both local and imported fish species, a combination of five or six fish species is used in a single fish pond. These species are selected so that they do not compete for food among them having different types of food habitats. As a result the food available in all the parts of the pond is used. Fish used in this system include catla and silver carp which are surface feeders, rohu a column feeder and mrigal and common carp which are bottom feeders. Other fish will also feed on the excreta of the common carp and this helps contribute to the efficiency of the system which in optimal conditions will produce 3000–6000 kg of fish per hectare per year.

## Integrated Recycling Systems

One of the largest problems with freshwater aquaculture is that it can use a million gallons of water per acre (about 1 $m^3$ of water per $m^2$) each year. Extended water purification systems allow for the reuse (recycling) of local water.

The largest-scale pure fish farms use a system derived (admittedly much refined) from the New Alchemy Institute in the 1970s. Basically, large plastic fish tanks are placed in a greenhouse. A hydroponic bed is placed near, above or between them. When tilapia are raised in the tanks, they are able to eat algae, which naturally grows in the tanks when the tanks are properly fertilized.

The tank water is slowly circulated to the hydroponic beds where the tilapia waste feeds commercial plant crops. Carefully cultured microorganisms in the hydroponic bed convert ammonia to nitrates, and the plants are fertilized by the nitrates and phosphates. Other wastes are strained out by the hydroponic media, which doubles as an aerated pebble-bed filter.

This system, properly tuned, produces more edible protein per unit area than any other. A wide variety of plants can grow well in the hydroponic beds. Most growers concentrate on herbs (e.g. parsley and basil), which command premium prices in small quantities all year long. The most common customers are restaurant wholesalers.

Since the system lives in a greenhouse, it adapts to almost all temperate climates, and may also adapt to tropical climates. The main environmental impact is discharge of water that must be salted to maintain the fishes' electrolyte balance. Current growers use a variety of proprietary tricks to keep fish healthy, reducing their expenses for salt and waste water discharge permits. Some veterinary authorities speculate that ultraviolet ozone disinfectant systems (widely used for ornamental fish) may play a prominent part in keeping the Tilapia healthy with recirculated water.

A number of large, well-capitalized ventures in this area have failed. Managing both the biology and markets is complicated.

## Classic Fry Farming

This is also called a "Flow through system" Trout and other sport fish are often raised from eggs to fry or fingerlings and then trucked to streams and released. Normally, the fry are raised in long, shallow concrete tanks, fed with fresh stream water. The fry receive commercial fish food in pellets. While not as efficient as the New Alchemists' method, it is also far simpler, and has been used for many years to stock streams with sport fish. European eel (Anguilla anguilla) aquaculturalists procure a limited supply of glass eels, juvenile stages of the European eel which swim north from the Sargasso Sea breeding grounds, for their farms. The European eel is threatened with extinction because of the excessive catch of glass eels by Spanish fishermen and overfishing of adult eels in, e.g., the Dutch IJsselmeer, Netherlands. As per 2005, no one has managed to breed the European eel in captivity.

## ISSUES

The issue of feeds in fish farming has been a controversial one. Many cultured fishes (tilapia, carp, catfish, many others) require no meat or fish products in their diets. Top-level carnivores (most salmon species) depend on fish feed of which a portion is usually derived from wild caught (anchovies, menhaden, etc.). Vegetable-derived proteins have successfully replaced fish meal in feeds for carnivorous fishes, but vegetable-derived oils have not successfully been incorporated into the diets of carnivores.

Secondly, farmed fish are kept in concentrations never seen in the wild (e.g. 50,000 fish in a 2-acre (8,100 m2) area.) with each fish occupying less room than the average bathtub. This can cause several forms of pollution. Packed tightly, fish rub against each other and the sides of their cages, damaging their fins and tails and becoming sickened with various diseases and infections. This also causes stress.

However, fish tend also to be animals that aggregate into large schools at high density. Most successful aquaculture species are schooling species, which do not have social problems at high density. Aquaculturists tend to feel that operating a rearing system above its design capacity or above the social density limit of the fish will result in decreased growth rate and increased FCR (food conversion ratio - kg dry feed/kg of fish produced), which will result in increased cost and risk of health problems along with a decrease in profits. Stressing the animals is not desirable, but the concept of and measurement of stress must be viewed from the perspective of the animal using the scientific method.

Sea lice, particularly *Lepeophtheirus salmonis* and various *Caligus* species, including *Caligus clemensi* and *Caligus rogercresseyi*, can cause deadly infestations of both farm-grown and wild salmon. Sea lice are ectoparasites which feed on mucus, blood, and skin, and migrate and latch onto the skin of wild salmon during free-swimming, planktonic *nauplii* and *copepodid* larval stages, which can persist for several days. Large numbers of highly populated, open-net salmon farms can create exceptionally large concentrations of sea lice; when exposed in river estuaries containing large numbers of open-net farms, many young wild salmon are infected, and do not survive as a result. Adult salmon may survive otherwise critical numbers of sea lice, but small, thin-skinned juvenile salmon migrating to sea are highly vulnerable. On the Pacific coast of Canada, the louse-induced mortality of pink salmon in some regions is commonly over 80%.

A 2008 meta-analysis of available data shows that salmon farming reduces the survival of associated wild salmon populations. This relationship has been shown to hold for Atlantic, steelhead, pink, chum, and coho salmon. The decrease in survival or abundance often exceeds 50 percent.

Diseases and parasites are the most commonly cited reasons for such decreases. Some species of sea lice have been noted to target farmed coho and Atlantic salmon. Such parasites have been shown to have an effect on nearby wild fish. One place that has garnered international media attention is British Columbia's Broughton

Archipelago. There, juvenile wild salmon must "run a gauntlet" of large fish farms located off-shore near river outlets before making their way to sea. It is alleged that the farms cause such severe sea lice infestations that one study predicted in 2007 a 99% collapse in the wild salmon population by 2011. This claim, however, has been criticized by numerous scientists who question the correlation between increased fish farming and increases in sea lice infestation among wild salmon.

Because of parasite problems, some aquaculture operators frequently use strong antibiotic drugs to keep the fish alive (but many fish still die prematurely at rates of up to 30 percent). In some cases, these drugs have entered the environment. Additionally, the residual presence of these drugs in human food products has become controversial. Use of antibiotics in food production is thought to increase the prevalence of antibiotic resistance in human diseases. At some facilities, the use of antibiotic drugs in aquaculture has decreased considerably due to vaccinations and other techniques. However, most fish farming operations still use antibiotics, many of which escape into the surrounding environment.

The lice and pathogen problems of the 1990s facilitated the development of current treatment methods for sea lice and pathogens. These developments reduced the stress from parasite/pathogen problems. However, being in an ocean environment, the transfer of disease organisms from the wild fish to the aquaculture fish is an ever-present risk.

The very large number of fish kept long-term in a single location contributes to habitat destruction of the nearby areas. The high concentrations of fish produce a significant amount of condensed faeces, often contaminated with drugs, which again affect local waterways. However, these effects are very local to the actual fish farm site and are minimal to non-measurable in high current sites.

Concern remains that resultant bacterial growth strips the water of oxygen, reducing or killing off the local marine life. Once an area has been so contaminated, the fish farms are moved to new, uncontaminated areas. This practice has angered nearby fishermen.

Other potential problems faced by aquaculturists are the obtaining of various permits and water-use rights, profitability, concerns about invasive species and genetic engineering depending on what species are involved, and interaction with the United Nations Convention on the Law of the Sea.

In regards to genetically modified farmed salmon, concern has been raised over their proven reproductive advantage and how it could potentially decimate local fish populations, if released into the wild. Biologist Rick Howard did a controlled laboratory study where wild fish and GMO fish were allowed to breed. The GMO fish crowded out the wild fish in spawning beds, but the offspring were less likely to survive. The colorant used to make pen-raised salmon appear rosy like their wild cousins has been linked with retinal problems in humans.

## Labeling

In 2005, Alaska passed legislation requiring that any genetically altered fish sold in the state be labeled. In 2006, a Consumer Reports investigation revealed that farm-raised salmon is frequently sold as wild.

In 2008, the US National Organic Standards Board allowed farmed fish to be labeled as organic provided less than 25% of their feed came from wild fish. This decision was criticized by the advocacy group Food & Water Watch as "bending the rules" about organic labeling. In the European Union, fish labeling as to species, method of production and origin, has been required since 2002.

Concerns continue over the labeling of salmon as farmed or wild caught, as well as about the humane treatment of farmed fish. The Marine Stewardship Council has established an Eco label to distinguish between farmed and wild caught salmon, while the RSPCA has established the Freedom Food label to indicate humane treatment of farmed salmon as well as other food products.

# INDOOR FISH FARMING

An alternative to outdoor open ocean cage aquaculture, is through the use of a recirculation aquaculture system (RAS). A RAS is a series of culture tanks and filters where water is continuously recycled and monitored to keep optimal conditions year round. To prevent the deterioration of water quality, the water is treated mechanically through the removal of particulate matter and biologically through the conversion of harmful accumulated chemicals into nontoxic ones.

Other treatments such as UV sterilization, ozonation, and oxygen injection are also used to maintain optimal water quality. Through this system, many of the environmental drawbacks of aquaculture are minimized including escaped fish, water usage, and the introduction of pollutants. The practices also increased feed-use efficiency growth by providing optimum water quality (Timmons et al., 2002; Piedrahita, 2003).

One of the drawbacks to recirculation aquaculture systems is water exchange. However, the rate of water exchange can be reduced through aquaponics, such as the incorporation of hydroponically grown plants (Corpron and Armstrong, 1983) and denitrification (Klas et al., 2006). Both methods reduce the amount of nitrate in the water, and can potentially eliminate the need for water exchanges, closing the aquaculture system from the environment. The amount of interaction between the aquaculture system and the environment can be measured through the cumulative feed burden (CFB kg/M3), which measures the amount of feed that goes into the RAS relative to the amount of water and waste discharged.

Because of its high capital and operating costs, RAS has generally been restricted to practices such as broodstock maturation, larval rearing, fingerling production, research animal production, SPF (specific pathogen free) animal production, and caviar and ornamental fish production. Although the use of RAS for other species is considered by many aquaculturalists to be impractical, there has been some limited successful implementation of this with high value product such as barramundi, sturgeon and live tilapia in the US.

## SLAUGHTER METHODS

Tanks saturated with carbon dioxide have been used to make fish unconscious. Then their gills are cut with a knife so that the fish bleed out before they are further processed. This is no longer considered a humane method of slaughter. Methods that induce much less physiological stress are electrical or percussive stunning and this has led to the phasing out of the carbon dioxide slaughter method in Europe.

### Inhumane Methods

According to T. Håstein of the National Veterinary Institute, "Different methods for slaughter of fish are in place and it is no doubt that many of them may be considered as appalling from an animal welfare point of view." A 2004 report by the EFSA Scientific Panel on Animal Health and Welfare explained: "Many existing commercial killing methods expose fish to substantial suffering over a prolonged period of time. For some species, existing methods, whilst capable of killing fish humanely, are not doing so because operators don't have the knowledge to evaluate them." Following are some of the less humane ways of killing fish.

- Air Asphyxiation. This amounts to suffocation in the open air. The process can take upwards of 15 minutes to induce death, although unconsciousness typically sets in sooner.
- Ice baths / chilling. Farmed fish are sometimes chilled on ice or submerged in near-freezing water. The purpose is to dampen muscle movements by the fish and to delay the onset of post-death decay. However, it does not necessarily reduce sensibility to pain; indeed, the chilling process has been shown to elevate cortisol. In addition, reduced body temperature extends the time before fish lose consciousness.
- $CO_2$ narcosis.
- Exsanguination without stunning. This is a process in which fish are taken up from water, held still, and cut so as to cause bleeding. According to references in Yue , this can leave fish writhing for an average of four minutes, and some catfish still responded to noxious stimuli after more than 15 minutes.

### More Humane Methods

- Percussive stunning.
- Electric stunning. This can be humane when a proper current, duration, conductivity, and temperature are present. One advantage is that in-water stunning allows fish to be rendered unconscious without stressful handling or displacement. However, improper stunning may not induce insensibility long enough to prevent the fish from enduring exsanguination while conscious. It's unknown whether the optimal stunning parameters that researchers have determined in studies are used by the industry in practice.

# AQUACULTURE

Aquaculture, also known as aquafarming, is the farming of aquatic organisms such as fish, crustaceans, molluscs and aquatic plants. Aquaculture involves cultivating freshwater and saltwater populations under controlled conditions, and can be contrasted with commercial fishing, which is the harvesting of wild fish. Mariculture refers to aquaculture practiced in marine environments.

The reported output from global aquaculture operations would supply one half of the fish and shellfish that is directly consumed by humans; however, there are issues about the reliability of the reported figures. Further, in current aquaculture practice, products from several pounds of wild fish are used to produce one pound of a piscivorous fish like salmon.

Particular kinds of aquaculture include fish farming, shrimp farming, oyster farming, algaculture (such as seaweed farming), and the cultivation of ornamental fish. Particular methods include aquaponics and Integrated multi-trophic aquaculture, both of which integrate fish farming and plant farming.

## History

The indigenous Gunditjmara people in Victoria, Australia may have raised eels as early as 6000 BC. There is evidence that they developed about 100 square kilometres (39 sq mi) of volcanic floodplains in the vicinity of Lake Condah into a complex of channels and dams, that they used woven traps to capture eels, and that capturing and smoking eels supported them year round.

Aquaculture was operating in China circa 2500 BC. When the waters subsided after river floods, some fishes, mainly carp, were trapped in lakes. Early aquaculturists fed their brood using nymphs and silkworm feces, and ate them. A fortunate genetic mutation of carp led to the emergence of goldfish during the Tang Dynasty.

Japanese cultivated seaweed by providing bamboo poles and, later, nets and oyster shells to serve as anchoring surfaces for spores.

## Romans bred fish in ponds.

In central Europe, early Christian monasteries adopted Roman aquacultural practices. Aquaculture spread in Europe during the Middle Ages, since away from the seacoasts and the big rivers, fish were scarce/expensive. Improvements in transportation during the 19th century made fish easily available and inexpensive, even in inland areas, making aquaculture less popular.

Hawaiians constructed oceanic fish ponds (see Hawaiian aquaculture). A remarkable example is a fish pond dating from at least 1,000 years ago, at Alekoko. Legend says that it was constructed by the mythical Menehune dwarf people.

In 1859 Stephen Ainsworth of West Bloomfield, New York, began experiments with brook trout. By 1864 Seth Green had established a commercial fish hatching operation at Caledonia Springs, near Rochester, New York. By 1866, with the involvement of Dr. W. W. Fletcher of Concord, Massachusetts, artificial fish hatcheries were under way in both Canada and the United States. When the Dildo Island fish hatchery opened in Newfoundland in 1889, it was the largest and most advanced in the world.

Californians harvested wild kelp and attempted to manage supply circa 1900, later labeling it a wartime resource.

## Definition

According to the FAO, aquaculture "is understood to mean the farming of aquatic organisms including fish, molluscs, crustaceans and aquatic plants. Farming implies some form of intervention in the rearing process to enhance production, such as regular stocking, feeding, protection from predators, etc. Farming also implies individual or corporate ownership of the stock being cultivated."

## 21st Century Practice

About 430 (97%) of the species cultured as of 2007 were domesticated during the 20th century, of which an estimated 106 came in the decade to 2007. Given the long-term importance of agriculture, it is interesting to note that to date only 0.08% of known land plant species and 0.0002% of known land animal species have been domesticated, compared with 0.17% of known marine plant species and 0.13% of known marine animal species. Domestication typically involves about a decade of scientific research. Domesticating aquatic species involves fewer risks to humans than land animals, which took a large toll in human lives. Most major human diseases originated in domesticated animals, through diseases such as smallpox and diphtheria, that like most infectious diseases, move to humans from animals. No human pathogens of comparable virulence have yet emerged from marine species.

Harvest stagnation in wild fisheries and overexploitation of popular marine species, combined with a growing demand for high quality protein, encourage aquaculturists to domesticate other marine species.

## Production Volume

In 2004, the total world production of fisheries was 140 million tonnes of which aquaculture contributed 45 million tonnes, about one third. The growth rate of worldwide aquaculture has been sustained and rapid, averaging about 8 percent per annum for over thirty years, while the take from wild fisheries has been essentially flat for the last decade. The aquaculture market reached $86 billion in 2009. Aquaculture is an especially important economic activity in China. Between 1980 and 1997, the Chinese Bureau of Fisheries reports, aquaculture harvests grew at an annual rate of 16.7 percent, jumping from 1.9 million tonnes to nearly 23 million tonnes. In 2005, China accounted for 70% of world production. Aquaculture is also currently one of the fastest growing areas of food production in the U.S.

Approximately 90% of all U.S. shrimp consumption is farmed and imported. In recent years salmon aquaculture has become a major export in southern Chile, especially in Puerto Montt, Chile's fastest-growing city.

### Over Reporting

China overwhelmingly dominates the world in reported aquaculture output. They report a total output which is double that of the rest of the world put together. However, there are issues with the accuracy of China's returns.

In 2001, the fisheries scientists Reg Watson and Daniel Pauly expressed concerns in a letter to *Nature*, that China was over reporting its catch from wild fisheries in the 1990s. They said that made it appear that the global catch since 1988 was increasing annually by 300,000 tonnes, whereas it was really shrinking annually by 350,000 tonnes. Watson and Pauly suggested this may be related to China policies where state entities that monitor the economy are also tasked with increasing output. Also, until recently, the promotion of Chinese officials was based on production increases from their own areas.

China disputes this claim. The official Xinhua News Agency quoted Yang Jian, director general of the Agriculture Ministry's Bureau of Fisheries, as saying that China's figures were "basically correct". However, the FAO accepts there are issues with the reliability of China's statistical returns, and currently treats data from China, including the aquaculture data, apart from the rest of the world.

## METHODS

### Mariculture

Mariculture is the term used for the cultivation of marine organisms in seawater, usually in sheltered coastal waters. In particular, the farming of marine fish is an example of mariculture, and so also is the farming of marine crustaceans (such as shrimps), molluscs (such as oysters) and seaweed.

### Integrated

Integrated Multi-Trophic Aquaculture (IMTA) is a practice in which the by-products (wastes) from one species are recycled to become inputs (fertilizers, food) for another. Fed aquaculture (for example, fish, shrimp) is combined with inorganic extractive (for example, seaweed) and organic extractive (for example, shellfish) aquaculture to create balanced systems for environmental sustainability (biomitigation), economic stability (product diversification and risk reduction) and social acceptability (better management practices).

"Multi-Trophic" refers to the incorporation of species from different trophic or nutritional levels in the same system. This is one potential distinction from the age-old practice of aquatic polyculture, which could simply be the co-culture of different fish species from the same trophic level. In this case, these organisms may all share the same biological and chemical processes, with few synergistic benefits, which could potentially lead to significant shifts in the ecosystem. Some traditional polyculture systems may, in fact, incorporate a greater diversity of species, occupying several niches, as extensive cultures (low intensity, low management) within the same pond. The "Integrated" in IMTA refers to the more intensive cultivation of the different species in proximity of each other, connected by nutrient and energy transfer through water.

Ideally, the biological and chemical processes in an IMTA system should balance. This is achieved through the appropriate selection and proportions of different species providing different ecosystem functions. The co-cultured species are typically more than just biofilters; they are harvestable crops of commercial value. A working IMTA system can result in greater total production based on mutual benefits to the co-cultured species and improved ecosystem health, even if the production of individual species is lower than in a monoculture over a short term period.

Sometimes the term "Integrated Aquaculture" is used to describe the integration of monocultures through water transfer. For all intents and purposes however, the terms "IMTA" and "integrated aquaculture" differ only in their degree of descriptiveness. Aquaponics, fractionated aquaculture, IAAS (integrated agriculture-aquaculture systems), IPUAS (integrated peri-urban-aquaculture systems), and IFAS (integrated fisheries-aquaculture systems) are other variations of the IMTA concept.

### Netting Materials

Various materials, including nylon, polyester, polypropylene, polyethylene, plastic-coated welded wire, rubber, patented rope products (Spectra, Thorn-D, Dyneema), galvanized steel and copper are used for netting in aquaculture fish enclosures around the world. All of these materials are selected for a variety of reasons, including design feasibility, material strength, cost, and corrosion resistance.

Recently, copper alloys have become important netting materials in aquaculture because they are antimicrobial (i.e., they destroy bacteria, viruses, fungi, algae, and other microbes) and they therefore prevent biofouling (i.e., the undesirable accumulation, adhesion, and growth of microorganisms, plants, algae, tubeworms, barnacles, mollusks, and other organisms). By inhibiting microbial growth, copper alloy aquaculture cages avoid costly net changes that are necessary with other materials. The resistance of organism growth on copper alloy nets also provides a cleaner and healthier environment for farmed fish to grow and thrive.

## SPECIES GROUPS

### Fish

The farming of fish is the most common form of aquaculture. It involves raising fish commercially in tanks, ponds, or ocean enclosures, usually for food. A facility that releases juvenile fish into the wild for recreational fishing or to supplement a species' natural numbers is generally referred to as a fish hatchery. Fish species raised by fish farms include salmon, bigeye tuna, carp, tilapia, catfish and cod.

In the Mediterranean, young bluefin tuna are netted at sea and towed slowly towards the shore. They are then interned in offshore pens where they are further grown for the market. In 2009, researchers in Australia managed for the first time to coax tuna (Southern bluefin) to breed in landlocked tanks.

## Crustaceans

Commercial shrimp farming began in the 1970s, and production grew steeply thereafter. Global production reached more than 1.6 million tonnes in 2003, worth about 9 billion U.S. dollars. About 75% of farmed shrimp is produced in Asia, in particular in China and Thailand. The other 25% is produced mainly in Latin America, where Brazil is the largest producer. Thailand is the largest exporter.

Shrimp farming has changed from its traditional, small-scale form in Southeast Asia into a global industry. Technological advances have led to ever higher densities per unit area, and broodstock is shipped worldwide. Virtually all farmed shrimp are penaeids (i.e., shrimp of the family *Penaeidae*), and just two species of shrimp, the Pacific white shrimp and the giant tiger prawn, account for about 80% of all farmed shrimp. These industrial monocultures are very susceptible to disease, which has decimated shrimp populations across entire regions. Increasing ecological problems, repeated disease outbreaks, and pressure and criticism from both NGOs and consumer countries led to changes in the industry in the late 1990s and generally stronger regulations. In 1999, governments, industry representatives, and environmental organizations initiated a program aimed at developing and promoting more sustainable farming practices.

Freshwater prawn farming shares many characteristics with, including many problems with marine shrimp farming. Unique problems are introduced by the developmental life cycle of the main species, the giant river prawn.

The global annual production of freshwater prawns (excluding crayfish and crabs) in 2003 was about 280,000 tonnes of which China produced 180,000 tonnes followed by India and Thailand with 35,000 tonnes each. Additionally, China produced about 370,000 tonnes of Chinese river crab.

## Molluscs

Abalone farming began in the late 1950s and early 1960s in Japan and China. Since the mid-1990s, this industry has become increasingly successful. Over-fishing and poaching have reduced wild populations to the extent that farmed abalone now supplies most abalone meat.

## Echinoderms

Commercially harvested echinoderms include sea cucumbers and sea urchins. In China, sea cucumbers are farmed in artificial ponds as large as 1,000 acres (400 ha).

## Algae

Microalgae, also referred to as phytoplankton, microphytes, or planktonic algae constitute the majority of cultivated algae.

Macroalgae, commonly known as seaweed, also have many commercial and industrial uses, but due to their size and specific requirements, they are not easily cultivated on a large scale and are most often taken in the wild.

## Issues

Aquaculture can be more environmentally damaging than exploiting wild fisheries on a local area basis but has considerably less impact on the global environment on a per kg of production basis. Local concerns include waste handling, side-effects of antibiotics, competition between farmed and wild animals, and using other fish to feed more marketable carnivorous fish. However, research and commercial feed improvements during the 1990s & 2000s have lessened many of these.

Fish waste is organic and composed of nutrients necessary in all components of aquatic food webs. In-ocean aquaculture often produces much higher than normal fish waste concentrations. The waste collects on the ocean bottom, damaging or eliminating bottom-dwelling life. Waste can also decrease dissolved oxygen levels in the water column, putting further pressure on wild animals.[49]

## Fish oils

Tilapia from aquaculture has been shown to contain more fat and a much higher ratio of omega-6 to omega-3 oils.

### *Impacts on Wild Fish*

Salmon farming currently leads to a high demand for wild forage fish. Fish do not actually produce omega-3 fatty acids, but instead accumulate them from either consuming microalgae that produce these fatty acids, as is the case with forage fish like herring and sardines, or, as is the case with fatty predatory fish, like salmon, by eating prey fish that have accumulated omega-3 fatty acids from microalgae. To satisfy this requirement, more than 50 percent of the world fish oil production is fed to farmed salmon.

In addition, as carnivores, salmon require large nutritional intakes of protein, protein which is often supplied to them in the form of forage fish. Consequently, farmed salmon consume more wild fish than they generate as a final product. To produce one pound of farmed salmon, products from several pounds of wild fish are fed to them. As the salmon farming industry expands, it requires more wild forage fish for feed, at a time when seventy five percent of the worlds monitored fisheries are already near to or have exceeded their maximum sustainable yield. The industrial scale extraction of wild forage fish for salmon farming then impacts the survivability of the wild predator fish who rely on them for food.

Fish can escape from coastal pens, where they can interbreed with their wild counterparts, diluting wild genetic stocks. Escaped fish can become invasive, out competing native species.

## Coastal Ecosystems

Aquaculture is becoming a significant threat to coastal ecosystems. About 20 percent of mangrove forests have been destroyed since 1980, partly due to shrimp farming. An extended cost–benefit analysis of the total economic value of shrimp aquaculture built on mangrove ecosystems found that the external costs were much higher than the external benefits. Over four decades, 269,000 hectares (660,000 acres) of Indonesian mangroves have been converted to shrimp farms. Most of these farms are abandoned within a decade because of the toxin build-up and nutrient loss.

Salmon farms are typically sited in pristine coastal ecosystems which they then pollute. A farm with 200,000 salmon discharges more fecal waste than a city of 60,000 people. This waste is discharged directly into the surrounding aquatic environment, untreated, often containing antibiotics and pesticides." There is also an accumulation of heavy metals on the benthos (seafloor) near the salmon farms, particularly copper and zinc.

## Genetic Modification

Salmon have been genetically modified for faster growth, although they are not approved for commercial use, in the face of opposition. One study, in a laboratory setting, found that modified salmon mixed with their wild relatives were aggressive in competing, but ultimately failed.

# ANIMAL WELFARE

As with the farming of terrestrial animals, social attitudes influence the need for humane practices and regulations in farmed marine animals. Under the guidelines advised by the Farm Animal Welfare Council good animal welfare means both fitness and a sense of well being in the animal's physical and mental state. This can be defined by the Five Freedoms:

- Freedom from hunger & thirst
- Freedom from discomfort
- Freedom from pain, disease, or injury
- Freedom to express normal behaviour
- Freedom from fear and distress

However, the controversial issue in aquaculture is whether fish and farmed marine invertebrates are actually sentient, or have the perception and awareness to experience suffering. Although no evidence of this has been found in marine invertebrates, recent studies conclude that fish do have the necessary receptors (nociceptors) to sense noxious stimuli and so are likely to experience states of pain, fear and stress. Consequently, welfare in aquaculture is directed at vertebrates; finfish in particular.

## Common Welfare Concerns

Welfare in aquaculture can be impacted by a number of issues such as stocking densities, behavioural interactions, disease and parasitism. A major problem in determining the cause of impaired welfare is that these issues are often all interrelated and influence each other at different times.

Optimal stocking density is often defined by the carrying capacity of the stocked environment and the amount of individual space needed by the fish, which is very species specific. Although behavioural interactions such as shoaling may mean that high stocking densities are beneficial to some species, in many cultured species high stocking densities may be of concern. Crowding can constrain normal swimming behaviour, as well as increase aggressive and competitive behaviours such as cannibalism, feed competition, territoriality and dominance/subordination hierarchies. This potentially increases the risk of tissue damage due to abrasion from fish-to-fish contact or fish-to-cage contact. Fish can suffer reductions in food intake and food conversion efficiency. In addition, high stocking densities can result in water flow being insufficient, creating inadequate oxygen supply and waste product removal. Dissolved oxygen is essential for fish respiration and concentrations below critical levels can induce stress and even lead to asphyxiation. Ammonia, a nitrogen excretion product, is highly toxic to fish at accumulated levels, particularly when oxygen concentrations are low.

Many of these interactions and effects cause stress in the fish, which can be a major factor in facilitating fish disease. For many parasites, infestation depends on the host's degree of mobility, the density of the host population and vulnerability of the host's defence system. Sea lice are the primary parasitic problem for finfish in aquaculture, high numbers causing widespread skin erosion and haemorrhaging, gill congestion,and increased mucus production. There are also a number of prominent viral and bacterial pathogens that can have severe effects on internal organs and nervous systems.

## Improving Welfare

The key to improving welfare of marine cultured organisms is to reduce stress to a minimum, as prolonged or repeated stress can cause a range of adverse effects. Attempts to minimise stress can occur throughout the culture process. During grow out it is important to keep stocking densities at appropriate levels specific to each species, as well as separating size classes and grading to reduce aggressive behavioural interactions. Keeping nets and cages clean can assist positive water flow to reduce the risk of water degradation.

Not surprisingly disease and parasitism can have a major effect on fish welfare and it is important for farmers not only to manage infected stock but also to apply disease prevention measures. However, prevention methods, such as vaccination, can also induce stress because of the extra handling and injection. Other methods include adding antibiotics to feed, adding chemicals into water for treatment baths and biological control, such as using cleaner wrasse to remove lice from farmed salmon.

Many steps are involved in transport, including capture, food deprivation to reduce faecal contamination of transport water, transfer to transport vehicle via nets or pumps, plus transport and transfer to the delivery location. During transport water needs to be maintained to a high quality, with regulated temperature, sufficient oxygen and minimal waste products. In some cases anaesthetics may be used in small doses to calm fish before transport.

## PROSPECTS

Global wild fisheries are in decline, with valuable habitat such as estuaries in critical condition. The aquaculture or farming of piscivorous fish, like salmon, does not help the problem because they need to eat products from other fish, such as fish meal and fish oil. Studies have shown that salmon farming has major negative impacts on wild salmon, as well as the forage fish that need to be caught to feed them. Fish that are higher on the food chain are less efficient sources of food energy.

Apart from fish and shrimp, some aquaculture undertakings, such as seaweed and filter-feeding bivalve mollusks like oysters, clams, mussels and scallops, are relatively benign and even environmentally restorative. Filter-feeders filter pollutants as well as nutrients from the water, improving water quality. Seaweeds extract nutrients such as inorganic nitrogen and phosphorus directly from the water, and filter-feeding mollusks can extract nutrients as they feed on particulates, such as phytoplankton and detritus.

Some profitable aquaculture cooperatives promote sustainable practices. New methods lessen the risk of biological and chemical pollution through minimizing fish stress, fallowing netpens, and applying Integrated Pest Management. Vaccines are being used more and more to reduce antibiotic use for disease control.

Onshore recirculating aquaculture systems, facilities using polyculture techniques, and properly sited facilities (for example, offshore areas with strong currents) are examples of ways to manage negative environmental effects.

Recirculating aquaculture systems (RAS) recycle water by circulating it through filters to remove fish waste and food and then recirculating it back into the tanks. This saves water and the waste gathered can be used in compost or, in some cases, could even be treated and used on land. While RAS was developed with freshwater fish in mind, scientist associated with the Agricultural Research Service have found a way to rear saltwater fish using RAS in low-salinity waters. Although saltwater fish are raised in off-shore cages or caught with nets in water that typically has a salinity of 35 parts per thousand (ppt), scientists were able to produce healthy pompano, a saltwater fish, in tanks with a salinity of only 5 ppt. Commercializing low-salinity RAS are predicted to have positive environmental and economical effects.

Unwanted nutrients from the fish food would not be added to the ocean and the risk of transmitting diseases between wild and farm-raised fish would greatly be reduced. The price of expensive saltwater fish, such as the pompano and combia used in the experiments, would be reduced. However, before any of this can be done researchers must study every aspect of the fish's lifecycle, including the amount of ammonia and nitrate the fish will tolerate in the water, what to feed the fish during each stage of its lifecycle, the stocking rate that will produce the healthiest fish, etc.

# 2

# TAXONOMY AND ANATOMY OF FISH

The term "fish" most precisely describes any non-tetrapod craniate (i.e. an animal with a skull and in most cases a backbone) that has gills throughout life and whose limbs, if any, are in the shape of fins. Unlike groupings such as birds or mammals, fish are not a single clade but a paraphyletic collection of taxa, including hagfishes, lampreys, sharks and rays, ray-finned fish, coelacanths, and lungfish.

In most cases, fish are "cold-blooded", adjusting their body temperatures to vary as ambient temperatures change. Fish are abundant in most bodies of water and can be found in nearly all aquatic environments, from high mountain streams to the abyssal and even hadal depths of the deepest oceans (e.g., gulpers and anglerfish). Fish exhibit greater species diversity than any other class of vertebrates as 32,000 species of fish have been discovered so far.

Fish, especially as food, are an important resource worldwide. Commercial and subsistence fishers hunt fish in wild fisheries or farm them in ponds or in cages in the ocean. They are also caught by recreational fishers, kept as pets, raised by fish keepers, and exhibited in public aquaria. Fish have had a role in culture through the ages, serving as deities, religious symbols, and as the subjects of art, books and movies.

## DIVERSITY OF FISH

Fish range in size from the huge 16-metre (52 ft) whale shark to the tiny 8-millimetre (0.3 in) stout infant fish.

Fish are very diverse and are categorized in many ways. Although, most fish species have probably been discovered and described, about 250 new ones are still discovered every year. According to latest scientific survey, 32,000 species of fish had been described bill date now in 2011. That is more than the combined total of all other vertebrates: mammals, amphibians, reptiles and birds.

Indeed, lungfish and coelacanths are closer relatives of tetrapods (such as mammals, birds, amphibians, etc.) than of other fish such as ray-finned fish or sharks, so the last common ancestor of all fish is also an ancestor to tetrapods. As paraphyletic groups are no longer recognised in modern systematic biology, the use of the term "fish" as a biological group must be avoided.

Many types of aquatic animals commonly referred to as "fish" are not fish in the sense given above; examples include shellfish, cuttlefish, starfish, crayfish and jellyfish. In earlier times, even biologists did not make a distinction – sixteenth century natural historians classified also seals, whales, amphibians, crocodiles, even hippopotamuses, as well as a host of aquatic invertebrates, as fish. All mammals, including Cetaceans like Whales and Dolphins, are not fish. In some contexts, especially in aquaculture, the true fish are referred to as finfish to distinguish them from these other animals.

A typical fish is ectothermic, has a streamlined body for rapid swimming, extracts oxygen from water using gills or uses an accessory breathing organ to breathe atmospheric oxygen, has two sets of paired fins, usually one or two (rarely three) dorsal fins, an anal fin, and a tail fin, has jaws, has skin that is usually covered with scales, and lays eggs.

Each criterion has exceptions. Tuna, swordfish, and some species of sharks show some warm-blooded adaptations—they can heat their bodies significantly above ambient water temperature. Streamlining and swimming performance varies from fish such as tuna, salmon, and jacks that can cover 10–20 body-lengths per second to species such as eels and rays that swim no more than 0.5 body-lengths per second. Many groups of freshwater fish extract oxygen from the air as well as from the water using a variety of different structures.

Lungfish have paired lungs similar to those of tetrapods, gouramis have a structure called the labyrinth organ that performs a similar function, while many catfish, such as *Corydoras* extract oxygen via the intestine or stomach. Body shape and the arrangement of the fins is highly variable, covering such seemingly un-fishlike forms as seahorses, pufferfish, anglerfish, and gulpers.

Similarly, the surface of the skin may be naked (as in moray eels), or covered with scales of a variety of different types usually defined as placoid (typical of sharks and rays), cosmoid (fossil lungfish and coelacanths), ganoid (various fossil fish but also living gars and bichirs), cycloid, and ctenoid (these last two are found on most bony fish).

There are even fish that live mostly on land. Mudskippers feed and interact with one another on mudflats and go underwater to hide in their burrows. The catfish *Phreatobius cisternarum* lives in underground, phreatic habitats, and a relative lives in waterlogged leaf litter.

## CLASSES OF INDIAN FISH FAUNA

The Indian fish fauna is divided into two classes, viz., Chondrichthyes and Osteichthyes. The Chondrichthyes are represented by 131 species under 67 genera, 28 families and 10 Orders in the Indian region. The annual average landings of the Indian Chondrichthyes is 33,442 tonnes, of which, 15,537 tonnes come from the east coast and 17,605 tonnes come from the west coast and the rest come from the Andaman and Nicobar, and Lakshadeep Islands.

The Indian Osteichthyes are represented by 2,415 species belonging to 902 genera, 226 families and 30 orders, of which, five families, notably the family Parapsilorhynchidae are endemic to India. These small hillstream fishes include a single genus, viz., *Parapsilorhynchus* which contains three species. They occur in the

Western Ghats, Satpura mountains and the Bailadila range in Madhya Pradesh only. Further, the fishes of the family Psilorhynchidae with the only genus Psilorhynchus are also endemic to the Indian region. Other fishes endemic to India include the genus *Olytra* and the species *Horaichthys* setnai belonging to the families Olyridae and Horaichthyidae respectively.

The latter occur from the Gulf of Kutch to Trivandrum coast. The endemic fish families form 2.21 per cent of the total bony fish families of the Indian region. 223 endemic fish species are found in India, representing 8.75 per cent of the total fish species known from the Indian region and 128 monotypic genera of fishes found in India, representing 13.20 per cent of the genera of fishes known from the Indian region.

About 22000 species of fishes have been recorded in the world; of which, about eleven percent are found in Indian waters. Out of the 2200 species so far listed, 73 (3.32%) belong to the cold freshwater regime, 544 (24.73%) to the warm fresh waters domain, 143 (6.50%) to the brackish waters and 1440 (65.45%) to the marine ecosystem. Adequate protection of ecosystems is a necessary requirement for survival of all species and proper care is needed to overcome anthropogenic stresses. In the case of commercial species, rational exploitation is a pre-requisite for sustainability of the resources.

## Freshwater Fishes

There are about 450 families of freshwater fishes globally. Roughly 40 are represented in India (warm freshwater species). About 25 of these families contain commercially important species. Number of endemic species in warm water is about 544. Major warm water species are : Bagarius bagarius, Catla catla, Channa marulius, C. punctatus, C. striatus, Cirrhinus mrigala, Clarias batrachus, Heteropneustes fossilis, Labeo bata, L. calbasu, L. rohita, Aorichthys seenghala, Notopterus chitala, N. notopterus, Pangasius pangasius, Rita rita, Wallago attu.

Cyprinids (family: Cyprinidae), Live fish (family: Anabantidae, Clariidae, Channidae, Heteropneustidae), Cat fish (family: Bagridae, Silurdae, Schilbeidae), Clupeids (family: Clupeidae), Mullets (family: Mugilidae), featherbacks (family: Notopteridae), Loaches (family: Cobitidae), Eels (family: Mastacembelidae), Glass fishes (family: Chandidae) and Gobies (family: Gobiidae) are the major groups of fresh water fishes found in India.

Cyprinidae is one of the largest families and is well represented in India with species ranging from few milimeters in length (minnows) to more than a metre (major carps). Among the 544 freshwater fish species in India, Cyprinidae accounts for nearly 24.12% of them.

Many of the Cyprinids, especially carps, are widely captured and form the mainstay of culture operations. Depending on their growth and utility in culture systems, the carps are grouped as the major carps (Catla catla, Cirrhinus mrigala, Labeo calbasu, Labeo rohita) and minor carps (L. fimbriatus, L. bata, C. cirrhosus, C. reba). A few exotic carps, as indicated below, have also been introduced into the country mainly for culture puroposes (Hypopthalmichthys molitrix, Ctenopharyngodon idella, Cyprinus carpio). The mahseers are also included in cyprinidae. Though many species of the genus Tor are found in the high altitude cold waters, some species like Tor khudree and Tor putitora occur in warmer waters.

The genus *Tor* thus reveals a wide range of species diversity as regards adaptation to different ecological conditions. There is a serious decline of mahseer fishery in different ecosystems endangering its very existence, thus warranting adequate conservation of the fishes.

## Western Ghats : Hotspots of Biodiversity

The Western Ghats, one among the 25 global hotspots of biodiversity, harbours very rich flora and fauna. There are records of over 4,000 species of flowering plants (38% endemism), 330 butterflies (11% endemic), 156 reptiles (62% endemism), 508 birds (4% endemism), 120 mammals (12% endemism) (Daniels, 2003), 290 fishes (41% endemism); (Babu and Nayar, 2004) and 131 amphibians (75% endemism). The varied climate and diverse topography create a wide array of habitats that support unique sets of plant and animal species. The level of endemism is high and the region is considered one of the world's biodiversity hotspots.

## Fishes in Freshwater

Fishes are the most numerous vertebrates living on this earth and worldwide there are over 25000 species of fishes. Of this about 48% live in freshwaters that constitute just 0.01% of the earth's water. Freshwater fish diversity is unevenly distributed on this planet. The species richness is high in tropical region compared to other parts of the earth. Usually these regions are characterized by high levels of endemism.

The world's major rivers like Amazon, Congo, Nile, etc. are some of the pristine rivers of the world with respect to freshwater fish diversity.

It has been estimated that the river Amazon and its tributaries may together harbour 3000 or more species of fishes. Such species-rich areas are called 'hotspots' and dominate other patterns or trends.

Probably, the climatic conditions of the tropical region are more stable compared to the temperate regions of the world. This could be one of the favourable conditions for the growth, survival and evolution for the species in tropical regions.

While a great deal of attention has been given to the loss of biodiversity in tropical rain forests, or in coastal areas, the diversity of and within freshwaters has been widely neglected. There is little doubt that freshwater fishes represent the most threatened set of vertebrates. In classifying the worlds' top 25 biodiversity hotspots, vertebrate group was considered excluding fish. This is mainly because of the poorly available data wherein the author predicts that there could be at least 5,000 species waiting to be discovered among fish, which is more than all mammals.

## FRESHWATER FISH DIVERSITY OF WESTERN GHATS

Many attempts mainly focused on evolving with a comprehensive checklist of freshwater fishes, which is an out come of the patchy (may be of a river basin, a region in the Western Ghats, an administrative boundary within the Western Ghats, etc) taxonomic information available on the diversity of freshwater fishes.

Daniels (2001) has listed 218 species from the Western Ghats of which 114 (52%) are endemic to Western Ghats.

However, this report lacks a detailed checklist of fishes found in the Western Ghats. The subsequent checklist (Shaji *et al* 2001) listed 287 fishes with names of individual species. This compilation considered certain estuarine fishes that are found to ascend freshwater for longer distances. The list highlighted the presence of 67% endemic species and 18 exotic or transplanted to the region. The most recent information available is by Dahanukar *et al* 2004 that lists 288 freshwater fishes, of which 118 (41%) are endemic to Western Ghats.

The threat status of fishes found in Western Ghats suggests that at least 41% of fish fauna is threatened by either being vulnerable, endangered, or critically endangered. This study also necessitates the implication of potent conservation measures to conserve the fish fauna of Western Ghats.

### Present Scenario

Present compilation of the checklist of the freshwater fishes in Western Ghats region lists 318 species of which 42.8% (136 species) are endemic to the region. Of this about 27 species are critically endangered and 55 endangered while 128 are data deficient. Altogether, 39.1% (123 species) of the freshwater fishes come under the category of critically endangered, endangered and vulnerable. Of the 27 critically endangered species 24 are endemic to the region. Similarly, of the 55 endangered species, 37 are endemic. Yet 49 endemic species are data deficient. A comparison of IUCN status between endemic and non-endemic species, which clearly shows that the endemic species comprises more of threatened species and the non-endemic comprise more of generalist species in Western Ghats.

### Unique Habitats for Fishes

The Western Ghats gifted with perennial streams outlines the unique habitats for the fishes. They run for about 1600 km from north to south and are drained by 38 east flowing and 27 west flowing major rivers. It is now identified that these pristine habitats are unique biological regions and 'hotspots' of fish diversity. The river systems consist of mainly the *running water habitat* and *lacustrine* habitats. There are numerous small lakes and ponds throughout the Western Ghats, which form the standing water habitats. Apart from these, there are forest swamps, deep wells, estuaries, gorges, etc, which are responsible for diverse fish fauna of the region.

## Species Composition

It can be observed that the species composition in these different habitats differs significantly between each other. Hill-streams dominate with the species of Balitoridae, especially the species of the genus *Schistura*. The torrential habitats compose of the highly adapted species of *Garra, Glypththorax*, etc. Shallow gradually flowing streams are rich with most general and small species (Ex. *Puntius, Danio, Rasbora, Aplocheilus,* etc). Species composition in the lacustrine habitats is mainly influenced by human beings in terms of exotic species introduction and large-scale harvesting. In these types of habitats large and medium commercial fishes dominate along with native commercial fishes. The exotic fish in the Western Ghats are *Cyprinus carpio, Carassius carassius, Tinca tinca, Oreochromis mossambicus* and the rainbow trout *Oncorhynchus mykiss* (golden and ordinary strains). The estuaries are dominated with dadromous and secondary freshwater fishes. Some of the wells in Kerala are recorded with the blind catfishes such as *Horaglanis krishnai* and *H. alikunhi.*

## *Freshwater Ecosystem*

The freshwater ecosystems of the Western Ghats are characterized by their patchy nature. Due to which each freshwater ecosystem, say a river or lake can be considered as an island with its own environmental characteristics.

This situation might have allowed to occur allopatric speciation and several species are known only from a drainage or a portion of the drainage system. This could be one of the important reasons behind, the exceptional fish diversity in Western Ghats and yet so endemic and threatened.

Natural inter-basin transfer of fish fauna takes place when the terrain is flat and receives heavy rainfall that inundates the entire region. In case of Western Ghats, although rainfall is of the range of 6000 mm annually, undulating terrain and great slopes of the basin provides little scope for inter-basin transfer of the fish fauna. However, recent anthropogenic activities in the form of damming the rivers and inter-basin transfer of water are the possible pathway for fish fauna to exchange between river basins.

## Monsoon Season

These are the connections available for diadromous and estuarine fishes to move from one basin to another, during monsoon season, when large amount of freshwater enters the sea from the estuary. Fishes that can withstand the reduced salinity of the sea possibly migrate to the nearest basin and recolonise.

## Movement of Fishes

Within a river basin, presence of falls greatly hinders the upstream and downstream movement of fishes. Most of the west-flowing rivers in the Western Ghats originate at the hilltop and after covering certain distance, they leap down over a great depth, forming natural falls. A study carried out by Bhat (2003) in four west flowing rivers of the Western Ghats of Karnataka found that upper reaches of the falls between the rivers were more similar to each other than comparing upper and lower reaches on the same river.

## DEFORESTATION AND TRANSFORMATION

The Western Ghats region, like other parts of the tropics, is undergoing rapid transformation. The deforestation rate is high and forests are being transformed into agriculture and monoculture plantations. Hydroelectric projects, mining, and extraction of forest products are also altering the landscape. Major human induced ecological changes in the Western Ghats begin with the arrival of agriculture and pastoralism. A climatic change towards the middle of fourth millennium BP, which induced widespread human migrations within the Indian sub-continent, is correlated to decline of forests and mangroves in Western Ghats. The ecosystems in the Western Ghats do bear the imprints of human actions through history.

### Threat to Global Biodiversity

Fragmentation, the loss and isolation of natural habitats, is one of the greatest threats to regional and global biodiversity. Habitat fragmentation takes place when the continuity is broken. This may be mainly through the construction of dams across the rivers, conversion of wetlands into other forms of land uses, destruction of riparian vegetation, etc.

Eight kinds of species have been identified that are likely to be especially vulnerable to the effects of fragmentation: rare species, species with large home ranges, species with limited powers of dispersal, species with low reproductive potential, species with short life cycles, species dependent on resources that are unpredictable in time or space, species of habitat interiors, and species exploited or persecuted by people. Due to the terrain of the western ghats, the rivers are extensively harvested for hydroelectric power through the construction of dams resulting in the fragmentation of the rivers. Anadromous fishes begin their life in freshwater, move to the marine environment to grow and mature, and then return to freshwater to reproduce. In western ghats decline in the population of *Anguilla bengalensis* has been attributed to fragmented rivers.

### Land Use Change in the Catchment

The cumulative geographical, hydrological, and ecological impacts of deforestation, timber harvesting, destruction of riparian vegetation, agricultural development, urbanization, land drainage, and flood protection on rivers have been documented worldwide. Deforestation of the catchment area usually results in soil erosion and nutrient input into the rivers.

Destruction of riparian vegetation damages the coarse stream substrate where several fishes find their shelter. High suspended sediment concentrations in rivers and sediment deposition on the riverbed can affect the health, habitat and particularly damage the gills. It has been estimated that about 25.6 % of the forest cover in Western Ghats is lost over a span of 22 years.

The Western Ghats region is prone to large-scale mining activities. These activities contribute huge quantities of silt into the aquatic systems. The rapid transformation of habitats in the form of silt deposition may result in the total changes in the species composition of the particular habitat.

## Flow Regulation

It is mainly due to the construction of dams across the rivers. It can be estimated that more than 60 % of the worlds is and many rivers have reduced to cascades of manmade lakes. This clearly reveals the magnitude of flow regulations all over the world. The flow regulation removes the lotic habitats and give rise to lentic habitats. Also, the homogenization of the habitats and within which it forms the temperature gradients.

## Concept of Interlinking

The concept of interlinking of river basins is gaining importance in recent years. Several such schemes in India and abroad have already being commissioned. However the serious repercussions in terms of water balance, water quality, the spread of pests and diseases, alterations to natural biotic distribution patterns and the disruption of significant ecological processes. In Indian context, the Western Ghats are characterised by high levels of fish endemism and the Himalayan region in terms of high diversity. Interlinking of the rivers of India may disrupt the distribution patterns of the fishes. Some fishes may find new optimal habitats and flourish prolifically on the expense of other native fishes.

Possibly the native fishes may be rare and endemic fishes needing immediate conservation measures. Thus in the process several species may become extinct and several may flourish. The concept of interlinking of the rivers of India is in the process of implementation. In this connection the above-discussed possibilities may be quite relevant to be considered.

## Aquatic Ecosystem

Fishes are regarded as the rich sources of protein. Throughout the world, fishing has offered employment to thousands of families. Subsistence fishing in streams, rivers and tanks has been a source of living for weaker section of the society. All these aspects have resulted in commercialization of fishery. The annual commercial catch from the fisheries of the world is approaching 100 million t.

Such a tremendous rate of extraction from the aquatic ecosystem has adversely affected the fish base of the world. Human uses of freshwater resources have threatened the sustainability of these resources. Increase in the fishing population and the demand for fish has intensified the fishing activities. Fishermen are using the destructive methods of fishing to get higher yields. Illegal usage of small meshed fishing gears that damage the juveniles, is still in use. Fishing using chemical and herbal poisons like copper sulphate, bleaching powder, tree extracts damage the entire fish population of the aquatic body. Dynamiting is another destructive method by which fishermen easily collect huge quantity of fishes. Use of dragnets which sweep away the entire fish resource of a particular area including juveniles can be disastrous to both target and non-target fishes of the area.

## Reservoir and Fish Yield

A number of species get introduced into the waters, most of which are to the reservoir areas of dams with a view to increase the net fish yield of the reservoir. Of these introduced fishes, many have naturalised in the streams, ponds and tanks of the region. Species like *Oreochromis mossambica*, *Gambusia offinis*, *Poecilia reticulata,* etc. are exotic species, which are originally from Africa and South America and introduced extensively in India.

Species like *Labeo rohita, Catla catla, Cirhinus mrigala* have been introduced recently in to the reservoirs. Importantly *Cyprinus carpio* has been widely introduced into the reservoirs of Western Ghats. This kind of haphazard introduction not only jeopardized the reservoirs but also threatened the freshwater biodiversity.

## Poor Water Quality

Changes in water quality due to discharge of effluents from various industrial, domestic and agricultural sources have deteriorated water quality. The toxic elements in pollutants may be total, killing all the fish species present, or selective, destroying a few sensitive species or so altering the environment that some species are favoured and others not.

## Outbreak of Diseases

Several incidences of outbreak of the diseases such as EUS (Epizootic Ulcerative syndrome) have been witnessed in freshwater habitats of Western Ghats. These have affected the sensitive native fishes of the region. For instance, in the last decade, large-scale mortality of freshwater fish took place due to EUS in most parts of the Western Ghats and the country as well.

## Fish Communities

Roughly 318 species of fish, which have been identified, are undoubtedly the relics of the several hundreds of species that have ever existed in this region. The reason behind existence of these modern survivors may be presence of some of the excellent basins in good condition within the region and traditional conservation practices in terms of protected sites, temple ponds, etc. The traditional practices involve declaring a stretch of river or a pond as sacred and giving the fullest protection to the fishes dwelling in them. However, to conserve the existing fish resources of the region, several other management strategies need to be designed and implemented.

Properly maintained scientific data on the fish fauna of the water body, participatory approaches in conservation, educating the fishing communities, managing the terrestrial ecosystem in healthy condition, stopping the introduction of exotic species, and careful extraction,

## CONCEPT OF BIODIVERSITY

The term "biodiversity" refers to the diversity of all life. Biological diversity is one measure of how healthy an ecosystem is. Healthy environments support high diversity; stressed ecosystems do not. A diverse community is also a stable community with an abundance of genetic and structural variation available to deal with unexpected environmental change.

Populations of organisms with more genetic diversity are better able to cope with environmental change than are populations with low diversity. The latter can become so highly adapted to local conditions that any environmental disturbance may be enough to drive them to extinction. Therefore, diversity is critical in ensuring the continued existence of our natural world.

To bring the issue even closer to home, the diversity of plants and animals on earth provides us with a wealth of products that we have come to depend on in our daily lives. Food, medicine, clothing, the air we breathe, and, in fact, our very survival is dependent on the life around us. Selfish human activity is resulting in a loss of biodiversity on a scale greater than the mass extinctions of 65 million years ago when the dinosaurs went extinct.

The recent rate of extinction due to habitat destruction has been estimated at 10,000 times the baseline rate (i.e., excluding major extinction events) prior to human influence. Thousands of species are lost each year, and we have no idea what tremendous benefits to mankind these organisms might have provided-a cure for cancer or an end to global hunger? By preserving biodiversity, we preserve a higher quality of life and, indeed, promote our continued existence.

### Preserving Biodiversity

The vast majority of the planet's biodiversity is made up of small, inconspicuous, and often microscopic organisms. Small organisms occupy less space, consume less food individually, complete their life cycles more quickly, and are able to partition the environment into more niches. More niches mean more species. Preserving biodiversity, then, means protecting small things, even when we may never see them.

This is the case with many of our inconspicuous, non-game fishes. They form the majority of our provincial fish diversity; so they are worthy of protection for this reason alone. However, they also perform a critical role in maintaining the health and well-being of our more "charismatic" fish species.

Many people, including anglers and naturalists, are familiar with a handful of the game fishes that occur in Alberta. Of this handful, a few, like northern pike, walleye, and sturgeon, are relatively easy to identify; others, like some of the trout species, require a little more attention to detail in order to recognize them at all life stages. Nevertheless, these species show up regularly in angler's creels, on supermarket shelves, in popular books, and even on TV; so most people are aware that they exist.

But what about the fishes that anglers don't catch regularly, or those that are of little obvious commercial value? In Alberta, these make up the majority of our fish fauna. Only about 18 of the approximately 60 species that occur here are considered game fishes.

The other 42 species, or 70% of Alberta's fishes, are relatively unknown to most people. Yet these small but often abundant species are critical components of aquatic ecosystems and the fundamental building blocks of food webs that support the large, predatory game fishes.

Predator-prey interactions among these species are the simplest and most easily understood concepts relating forage fishes to game species. Knowing "who eats whom" is important to an understanding of how ecosystems function as a whole. However, the linkages among fish species and their environment are far more intricate than a simple food chain.

Some species compete with each other for food and space, some parasitize their neighbours, some live in tight symbiotic relationships where each is completely dependent on the other. Interactions among species affect behaviours, life cycles, and life history strategies in such a way as to minimize niche overlap in co-existing species and allow all to survive together.

Ecosystems have been described as a "house of cards" where each card, or species, supports the others. Removing one piece can cause the entire structure to collapse. We often describe a species that is known to support many other species in a community as a keystone species.

These may not be the showy, charismatic organisms that we all recognize, but their importance in maintaining a healthy environment cannot be overstated. We need them more than many people realize. We must ensure they are protected from human disturbance-just as we protect the more conspicuous plants and animals. And the first step toward protection and conservation is to have people become more familiar with these creatures and to recognize their importance in the natural world.

## OVERFISHING: A THREAT TO MARINE BIODIVERSITY

Despite its crucial importance for the survival of humanity, marine biodiversity is in ever-greater danger, with the depletion of fisheries among biggest concerns.

Fishing is central to the livelihood and food security of 200 million people, especially in the developing world, while one of five people on this planet depends on fish as the primary source of protein. According to UN agencies, aquaculture - the farming and stocking of aquatic organisms including fish, molluscs, crustaceans and aquatic plants - is growing more rapidly than all other animal food producing sectors.

But amid facts and figures about aquaculture's soaring worldwide production rates, other, more sobering, statistics reveal that global main marine fish stocks are in jeopardy, increasingly pressured by overfishing and environmental degradation.

"Overfishing cannot continue," warned Nitin Desai, Secretary General of the 2002 World Summit on Sustainable Development, which took place in Johannesburg. "The depletion of fisheries poses a major threat to the food supply of millions of people." The Johannesburg Plan of Imple-mentation calls for the establishment of Marine Protected Areas (MPAs), which many experts believe may hold the key to conserving and boosting fish stocks. Yet, according to the UN Environment Programme's (UNEP) World Conservation Monitoring Centre, in Cambridge, UK, less than one per cent of the world's oceans and seas are currently in MPAs.

## Problem of Overfishing

The magnitude of the problem of overfishing is often overlooked, given the competing claims of deforestation, desertification, energy resource exploitation and other biodiversity depletion dilemmas. The rapid growth in demand for fish and fish products is leading to fish prices increasing faster than prices of meat. As a result, fisheries investments have become more attractive to both entrepreneurs and governments, much to the detriment of small-scale fishing and fishing communities all over the world. In the last decade, in the north Atlantic region, commercial fish populations of cod, hake, haddock and flounder have fallen by as much as 95%, prompting calls for urgent measures. Some are even recommending zero catches to allow for regeneration of stocks, much to the ire of the fishing industry.

According to a Food and Agriculture Organization (FAO) estimate, over 70% of the world's fish species are either fully exploited or depleted. The dramatic increase of destructive fishing techniques worldwide destroys marine mammals and entire ecosystems. FAO reports that illegal, unreported and unregulated fishing worldwide appears to be increasing as fishermen seek to avoid stricter rules in many places in response to shrinking catches and declining fish stocks.

Few, if any, developing countries and only a limited number of developed ones are on track to put into effect by this year the International Plan of Action to Prevent, Deter and Eliminate Unreported and Unregulated Fishing. Despite that fact that each region has its Regional Sea Conventions, and some 108 governments and the European Commission have adopted the UNEP Global Programme of Action for the Protection of the Marine Environment from Land based Activities, oceans are cleared at twice the rate of forests.

The Johannesburg forum stressed the importance of restoring depleted fisheries and acknowledged that sustainable fishing requires partnerships by and between governments, fishermen, communities and industry. It urged countries to ratify the Convention on the Law of the Sea and other instruments that promote maritime safety and protect the environment from marine pollution and environmental damage by ships. Only a multilateral approach can counterbalance the rate of depletion of the world's fisheries which has increased more than four times in the past four decades.

## Fish Taxonomy

Like all of the world's organisms, aquarium fish are named and organised into a scientific order that is generally accepted as being correct.

This classification system (taxonomy) and naming conventions (nomenclature) are sometimes open to debate, but the general framework is well established and widely recognised.

It is fair to think of the system as a family tree where closely related organisms are found on the same branches and more distant relatives are found further away on different branches.

Fish are a paraphyletic group: that is, any clade containing all fish also contains the tetrapods, which are not fish. For this reason, groups such as the "Class Pisces" seen in older reference works are no longer used in formal classifications.

Fish are classified into the following major groups:

- Class Myxini (hagfish)
- Class Pteraspidomorphi † (early jawless fish)
- Class Thelodonti †
- Class Anaspida †
- Class Petromyzontida or Hyperoartia
  (a) Petromyzontidae (lampreys)
- Class Conodonta (conodonts) †
- Class Cephalaspidomorphi † (early jawless fish)
  (a) (unranked) Galeaspida †
  (b) (unranked) Pituriaspida †
  (c) (unranked) Osteostraci †
- Infraphylum Gnathostomata (jawed vertebrates)
  (a) Class Placodermi † (armoured fish)
  (b) Class Chondrichthyes (cartilaginous fish)
  (c) Class Acanthodii † (spiny sharks)
  (d) Superclass Osteichthyes (bony fish)
    (i) Class Actinopterygii (ray-finned fish)
    (ii) Subclass Chondrostei
    (iii) Order Acipenseriformes (sturgeons and paddlefishes)
    (iv) Order Polypteriformes (reedfishes and bichirs).
    (v) Subclass Neopterygii
    (vi) Infraclass Holostei (gars and bowfins)
    (vii) Infraclass Teleostei (many orders of common fish)
    (viii) Class Sarcopterygii (lobe-finned fish)
    (ix) Subclass Coelacanthimorpha (coelacanths)
    (x) Subclass Dipnoi (lungfish)

## † - Indicates Extinct Taxon

Some palaeontologists contend that because Conodonta are chordates, they are primitive fish. The position of hagfish in the phylum chordata is not settled.

Phylogenetic research in 1998 and 1999 supported the idea that the hagfish and the lampreys form a natural group, the Cyclostomata, that is a sister group of the Gnathostomata.

The various fish groups account for more than half of vertebrate species. There are almost 28,000 known extant species, of which almost 27,000 are bony fish, with 970 sharks, rays, and chimeras and about 108 hagfish and lampreys.

A third of these species fall within the nine largest families; from largest to smallest, these families are Cyprinidae, Gobiidae, Cichlidae, Characidae, Loricariidae, Balitoridae, Serranidae, Labridae, and Scorpaenidae. About 64 families are monotypic, containing only one species. The final total of extant species may grow to exceed 32,500.

## FISH ANATOMY

Fish are animals that are cold-blooded, have fins and a backbone. Most fish have scales and breathe with gills. There are about 22,000 species of fish that began evolving around 480 million years ago. The largemouth bass illustrated above has the typical torpedo-like (fusiform) shape associated with many fishes.

Fins are appendages used by the fish to maintain its position, move, steer and stop. They are either single fins along the centerline of the fish, such as the dorsal (back) fins, caudal (tail) fin and anal fin, or paired fins, which include the pectoral (chest) and pelvic (hip) fins. Fishes such as catfish have another fleshy lobe behind the dorsal fin, called an adipose (fat) fin that is not illustrated here. The dorsal and anal fins primarily help fish to not roll over onto their sides. The caudal fin is the main fin for propulsion to move the fish forward.

The paired fins assist with steering,stopping and hovering.Scales in most bony fishes (most freshwater fishes other than gar that have ganoid scales, and catfish which have no scales) are either ctenoid or cycloid. Ctenoid scales have jagged edges and cycloid have smooth rounded edges. Bass and most other fish with spines have ctenoid scales composed of connective tissue covered with calcium. Most fishes also have a very important mucus layer covering the body that helps prevent infection.

In many freshwater fishes the fins are supported by spines that are rigid and may be quite sharp thus playing a defensive role. Catfish have notably hard sharp fins that you should be wary of. The soft dorsal and caudal fins are composed of rays, as are portions of other fins. Rays are less rigid and frequently branched.The gills are the breathing apparatus of fish and are highly vascularized giving them their bright red cover. An operculum (gill cover) that is a flexible bony plate protects the sensitive gills. Water is "inhaled" through the mouth, passes over the gills and "exhaled" from beneath the operculum. Fish see through their eyes and can detect colour.

The eyes are rounder in fish than mammals because of the refractive index of water and focus is achieved by moving the lens in and out, not distorting it as in mammals. Paired nostrils, or nares, in fish are used to detect odours in water and can be quite sensitive. Eels and catfish have particularly well developed senses of smell.The mouth's shape is a good clue to what fish eat.

The larger it is the bigger the prey it can consume. Fish have a sense of taste and may sample items to taste them before swallowing if they are not obvious prey items. Most freshwater fishes in Florida are omnivorous (eating both plant and animal matter).

Some are primarily piscivorous (eating mostly other fish). The imported grass carp is one of the few large fishes that are primarily herbivorous (eating plants). Fish may or may not have teeth depending on the species. Fish like chain pickerel and gar have obvious canine-shaped teeth. Other fish have less obvious teeth, such as the cardiform teeth in catfish which feel like a roughened area at the front of the mouth, or vomerine teeth that are tiny patches of teeth, for example, in the roof of a striped bass's mouth.

Grass carp and other minnows have pharyngeal teeth modified from their gill arches for grinding that are located in the throat.The lateral line is a sensory organ consisting of fluid filled sacs with hair-like sensory apparatus that are open to the water through a series of pores (creating a line along the side of the fish).

The lateral line primarily senses water currents and pressure, and movement in the water.The vent is the external opening to digestive urinary and reproductive tracts. In most fish it is immediately in front of the anal fin.

Fish anatomy is primarily governed by the physical characteristics of water, which is much denser than air, holds a relatively small amount of dissolved oxygen, and absorbs more light than air does.

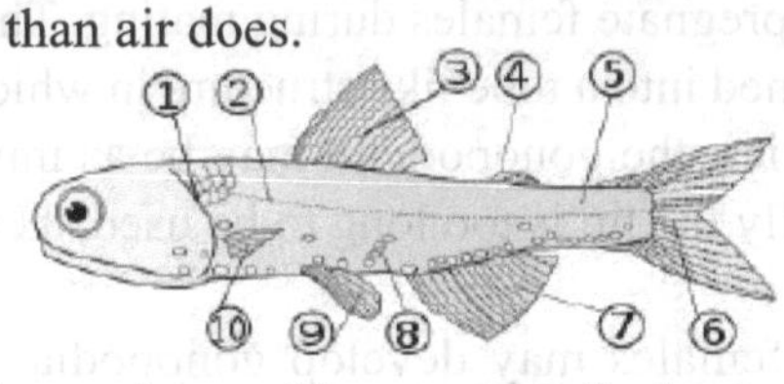


***The anatomy of Lampanyctodes hectoris***

## *Fins*

(1) - operculum (gill cover), (2) - lateral line, (3) - dorsal fin, (4) - adipose fin, — (5) - caudal peduncle, (6) - caudal fin, (7) - anal fin, (8) - photophores, — (9) - pelvic fins (paired), (10) - pectoral fins (paired)

The *fins* are the most distinctive features of a fish, composed of bony spines protruding from the body with skin covering them and joining them together, either in a webbed fashion, as seen in most bony fish, or more similar to a flipper, as seen in sharks. These usually serve as a means for the fish to swim. Fins can also be used for gliding or crawling, as seen in the flying fish and frogfish. Fins located in different places on the fish serve different purposes, such as moving forward, turning, and keeping an upright position.

## Spines Rays

In bony fish, most fins may have spines or rays. A fin may contain only spiny rays, only soft rays, or a combination of both. If both are present, the spiny rays are always anterior. Spines are generally stiff and sharp. Rays are generally soft, flexible, segmented, and may be branched. This segmentation of rays is the main difference that separates them from spines; spines may be flexible in certain species, but they will never be segmented.

Spines have a variety of uses. In catfish, they are used as a form of defense; many catfish have the ability to lock their spines outwards. Trigger fish also use spines to lock themselves in crevices to prevent them being pulled out.

Lepidotrichia are bony, bilaterally-paired, segmented fin rays found in bony fishes. They develop around actinotrichia as part of the dermal exoskeleton. Lepidotrichia may have some cartilage or bone in them as well. They are actually segmented and appear as a series of disks stacked one on top of another.

## Reproductive System

### *Internal Fertilization*

In many species of fish, fins have been modified to allow internal fertilization. A gonopodium is an anal fin that is modified into an intromittent organ in males of certain species of live-bearing fish in the families Anablepidae and Poeciliidae. It is movable and used to impregnate females during mating. The male's anal fin's 3rd, 4th and 5th rays are formed into a tube like structure in which the sperm of the fish is ejected. In some species, the gonopodium may be as much as 50% of the total body length. Occasionally the fin is too long to be used, as in the "lyretail" breeds of *Xiphophorus helleri*.

Hormone treated females may develop gonopodia. These are useless for breeding. One finds similar organs having the same characteristics in other types of fish, for example the andropodium in the *Hemirhamphodon* or in the Goodeidae.

When ready for mating, the gonopodium becomes "erect" and points forward, towards the female. The male shortly inserts the organ into the sex opening of the female, with hook-like adaptations that allow the fish to grip onto the female to ensure impregnation. If a female remains stationary and her partner contacts her vent with his gonopodium, she is fertilized. The sperm is preserved in the female's oviduct. This allows females to, at any time, fertilize themselves without further assistance of males.

Male cartilaginous fish have claspers modified from pelvic fins. These are intromittent organs, used to channel semen into the female's cloaca during copulation.

## Fish Scales

The outer body of many fish is covered with scales. Some species are covered instead by scutes. Others have no outer covering on the skin; these are called naked fish. Most fish are covered in a protective layer of slime (mucus).

There are four types of fish scales.

1. Placoid scales, also called dermal denticles, are similar to teeth in that they are made of dentin covered by enamel. They are typical of sharks and rays.
2. Ganoid scales are flat, basal-looking scales that cover a fish body with little overlapping. They are typical of gar and bichirs.
3. Cycloid scales are small oval-shaped scales with growth rings. Bowfin and remora have cycloid scales.
4. Ctenoid scales are similar to the cycloid scales, with growth rings. They are distinguished by spines that cover one edge. Halibut have this type of scale.

Another, less common, type of scale is the scute, which is:

- An external shield-like bony plate; or
- A modified, thickened scale that often is keeled or spiny, or
- A projecting, modified (rough and strongly ridged) scale, usually associated with the lateral line, or on the caudal peduncle forming caudal keels, or along the ventral profile. Some fish, such as pineconefish, are completely or partially covered in scutes.

## Vertebrae

The vertebrae of lobe-finned fishes consist of three discrete bony elements. The vertebral arch surrounds the spinal cord, and is of broadly similar form to that found in most other vertebrates. Just beneath the arch lies a small plate-like pleurocentrum, which protects the upper surface of the notochord, and below that, a larger arch-shaped intercentrum to protect the lower border. Both of these structures are embedded within a single cylindrical mass of cartilage. A similar arrangement was found in primitive tetrapods, but, in the evolutionary line that led to reptiles (and hence, also to mammals and birds), the intercentrum became partially or wholly replaced by an enlarged pleurocentrum, which in turn became the bony vertebral body.

In most ray-finned fishes, including all teleosts, these two structures are fused with, and embedded within, a solid piece of bone superficially resembling the vertebral body of mammals. In living amphibians, there is simply a cylindrical piece of bone below the vertebral arch, with no trace of the separate elements present in the early tetrapods.

In cartilagenous fish, such as sharks, the vertebrae consist of two cartilagenous tubes. The upper tube is formed from the vertebral arches, but also includes additional cartilagenous structures filling in the gaps between the vertebrae, and so enclosing the spinal cord in an essentially continuous sheath. The lower tube surrounds the notochord, and has a complex structure, often including multiple layers of calcification.

Lampreys have vertebral arches, but nothing resembling the vertebral bodies found in all higher vertebrates. Even the arches are discontinuous, consisting of separate pieces of arch-shaped cartilage around the spinal cord in most parts of the body, changing to long strips of cartilage above and below in the tail region. Hagfishes lack a true vertebral column, and are therefore not properly considered vertebrates, but a few tiny neural arches are present in the tail.

## The Jaw

Linkage systems are widely distributed in animals. The most thorough overview of the different types of linkages in animals has been provided by M. Muller, who also designed a new classification system, which is especially well suited for biological systems.

Linkage mechanisms are especially frequent and manifold in the head of bony fishes, such as wrasses, which have evolved many specialized feeding mechanisms. Especially advanced are the linkage mechanisms of jaw protrusion. For suction feeding a system of linked four-bar linkages is responsible for the coordinated opening of the mouth and 3-D expansion of the buccal cavity. Other linkages are responsible for protrusion of the premaxilla.

The vertebrate jaw probably originally evolved in the Silurian period and appeared in the Placoderm fish which further diversified in the Devonian. Jaws are thought to derive from the pharyngeal arches that support the gills in fish.

The two most anterior of these arches are thought to have become the jaw itself (see hyomandibula) and the hyoid arch, which braces the jaw against the braincase and increases mechanical efficiency. While there is no fossil evidence directly to support this theory, it makes sense in light of the numbers of pharyngeal arches that are visible in extant jawed (the Gnathostomes), which have seven arches, and primitive jawless vertebrates (the Agnatha), which have nine.

It is thought that the original selective advantage garnered by the jaw was not related to feeding, but to increased respiration efficiency. The jaws were used in the buccal pump (observable in modern fish and amphibians) that pumps water across the gills of fish or air into the lungs in the case of amphibians. Over evolutionary time, the more familiar use of jaws (to humans), in feeding, was selected for and became a very important function in vertebrates.

## Body Shapes

The way a fish looks is a good indicator of how it "makes a living." Body shape, mouth location and size, tail shape and color can reveal a lot about a fish's lifestyle.

Fusiform: Fusiform, or streamlined, fish like the barracuda or jack are capable of swimming very fast. They usually live in open water.

Laterally compressed: Fish that are laterally compressed (flattened from side to side) usually do not swim rapidly (some schooling fish are an exception). However, they are exceptionally maneuverable. Many, like the angelfish, are found near coral reefs. Their shape allows them to move about in the cracks and crevices of the reef. A flounder is a laterally compressed fish that lies on its side on the bottom. Both eyes migrate to the left or right side early in development.

Depressed: Depressed fish (flattened from top to bottom), like stingrays, live on the bottom.

Eel-like: Eel-like fish have a snake-like body shape. The electric eel and moral eels are good examples of fish with this body shape.

Many fish like the boxfish and porcupinefish do not fit into any of these categories. They are slow swimmers with special protective mechanisms.

## Tails

The shape of the tail can be an indicator of how fast a fish usually swims.

Crescent-shaped: Fish with crescent-shaped tails, like swordfish, are fast swimmers and constantly on the move.

Forked: Fish with forked tails, like the striped bass, are also fast swimmers, though they may not swim fast all of the time. The deeper the fork, the faster the fish can swim.

Rounded: Fish with a rounded or flattened tail are generally slow moving, but are capable of short, accurate bursts of speed. The mummichog has a tail with a rounded end.

## Mouths

The location and size of the mouth can be a good indicator of the food a fish eats and where it lives. Fish with large mouths generally eat large food items like another fish; however, the whale shark eats very small organisms which it strains from the water with its huge mouth. Fish with small mouths eat small food items: small crustaceans or molluscs; and, fish with tiny mouths eat tiny things like zooplankton.

Terminal: A terminal mouth is located on the end of the head. Fish with terminal mouths may chase and capture things, like the tuna, or pick at things, like the butterfly fish.

Up-Pointing: A fish with an up-pointing mouth has a long lower jaw. The mouth opening is toward the top of the head. The tarpon has this kind of mouth. It feeds near the surface.

Sub-Terminal: A sub-terminal mouth is on the underside of the head. Fish with this type of mouth usually feed on the bottom. The bonefish has a sub-terminal mouth.

Some fish have specialized mouths. The seahorse has a tiny mouth at the end of a straw-like snout that is used to “slurp”zooplankton.

## Color

Red is a common color in fish. You might think that red fish would be very easy for a predator to find. However, most fish that have this coloration live in dark or deep water, or are nocturnal (active at night). In deep water red light is filtered out quickly so red is a good camouflage. At night red-colored objects appear gray. The squirrelfish has this kind of coloration.

Countershading: Many fish are dark on top and light on the bottom. Most of these fish are found in open ocean. When seen from above they "disappear" by blending in with the dark color of the depths or the bottom. From below the light belly blends into the sky above. The tuna is countershaded. Seabirds like puffins are also countershaded.

Disruptive Coloration: This is a form of camouflage. The patterns and lines break up the outline of the fish or help it to blend into the background. The brightly patterned fish of coral reefs blend in with the corals despite their brilliant colors. The moorish idol exhibits disruptive coloration.

Eye Spots: Eye spots are a form of mimicry. The eye spot, usually found near the tail, draws attention away from the real eye which is a target that a predator might strike. The eye spot may cause the predator to attack the wrong end and allow the fish to escape alive.

Warning Coloration: Many fish use bright colors to "advertise" the presence of poisonous spines or some other defensive mechanism. The nave surgeonfish has two bAght orange spots near the base of the tail that advertise the presence of razor sharp spines.

Camouflage: Many fish have colors or patterns that match their backgrounds. The flounder is a camouflaged fish. It can even change color to match different backgrounds.

## HOW FISH GET OXYGEN

Fish need oxygen to survive. Since they do not have lungs they get their oxygen with gills. Oxygen in water passing across the membranes of the gills enters the fish's blood by diffusion. The oxygen that fish "breathe" is dissolved in the water. It enters the water at the surface by diffusion and as a byproduct of photosynthesis by aquatic plants.

Temperature and salinity are also important. Most fish are adapted to a narrow temperature range. Rapid temperature changes may cause death. Most fish are also restricted to a narrow range of salinity. Freshwater fish placed in salt water, or saltwater fish placed in fresh water may die. Some fish are adapted for life in estuaries where the salinity fluctuates and others, such as salmon or striped bass, are able to migrate between salt and fresh water.

## Five Senses

Fish have a number of senses that help them survive in their environments. Humans share some of them, but others are found only in fish.

Sight: Most fish have well developed eyes which are located on the side of the head. This positioning allows the fish to see in every direction. Fish that are colorful probably have color vision. Nocturnal fish have large eyes that help them see in the low light.

Smell and Taste: Fish can smell things in the water with two blind sacs called nares. Nares are similar to our nostrils, except fish cannot breathe through their nares. A fish can taste with taste buds in the lining of its mouth and gills. Some fish have feelers, like the "whiskers" on a catfish, called barbels, which are covered with taste buds.

Hearing: Fish have ears but you cannot see them. Their inner ears are well developed and pick up sound waves in the water through the fish's body.

Lateral Line: The lateral line system helps the fish feel movements in the water. The line, actually a row of tiny holes in the skin, begins behind the gill cover and runs along the side of the body to the tail. Tiny hairs in the lateral line system are sensitive to vibrations. This system helps fish swim in schools, avoid predators and find food.

Electricity: Some bony fish and sharks have special pores on the head that allow them to detect electrical currents. This sense aids them in navigating or finding prey in dark or muddy water.

## Fish Behavior

Behavior is the action of a fish in response to its environment including other animals. The most interesting and sometimes the most obvious behaviors involve interactions with other fish.

Migration: Many species of fish migrate during their life cycle. Salmon and some species of eel make long migrations to spawn. This reduces competition for food and space between adults and young fish. Other fish migrate to escape cold temperatures or to find food.

Schooling: Many fish gather in groups called schools. A school is a group of fish of the same size and species moving in the same direction. Schooling may increase the chances of finding food, avoiding predators and finding a mate. It may also make swimming easier. Sight is important in keeping the school together. The lateral line may also help.

Aggression: Aggressive behavior is an interaction between two fish of the same species or different species. It is usually associated with reproduction and defense of feeding territory. It includes posturing, direct attacks and ritualistic displays such as fin flaring and changes of color. During the breeding season some fish protect spawning territories.

Resting: Fish don't sleep like people, but some fish spend a good portion of their day or night resting. Coral reef fish that are active during the day spend the night in caves and crevices that nocturnal (night active) fish hide in during the day. This may reduce competition, conserve energy and keep the fish safe from predators.

Communication: Fish communicate with one another in many ways. Senses play a part in communication. Visual communication is important to most fish. Body movements, postures, colors, color patterns and light are the primary means of visual communication. Sound is also used for communication. Sounds are produced by grinding teeth, rubbing body parts together and with the swim bladder, an air sac inside the body used to regulate buoyancy.

Drum, croakers and toadfish get their names from the sounds they produce with their swim bladders. Fish also communicate by producing chemicals called pheromones that other fish can smell. A few fish that live in very muddy water where visual communication is impossible use electricity to communicate.

Cleaning: Some fish are cleaners. They pick parasites from other fish. Cleaning fish, such as wrasses, are brightly colored or patterned. They establish cleaning stations that other fish approach in order to be cleaned. Look for fish in a funny position or holding their fins out. They are signaling that they are ready to be cleaned.

Many of these behaviors can be observed during an Aquarium visit. Such behaviors indicate that the Husbandry staff has been successful in duplicating the fishes' environments.

## INTERNAL FISH ANATOMY

As different as a man may be from a fish, both creatures share some fascinating similarities in basic structure and function. And the closer one looks, the more complex life becomes. The smallest units of life are microscopic cells, and some organisms such as an ameba are no larger than a single cell. In larger multicellular creatures, individual cells that are similar in structure and perform a specific function are grouped into tissues, and tissues may be grouped into even more complex and specialized structures called organs.

These organs perform the basic bodily functions such as respiration, digestion, and sensory reception. Man and fish share such organs as the brain, stomach, liver, and kidneys. Other organs appear in different forms in different organisms; for example, the lungs in humans and the gills in fish are very different but both provide the same basic function of respiration.

Finally, some organs (such as the fish's swim bladder) are simply not present in man. Below are descriptions of some of the organs. A number of other vital organs, such as the spleen and pancreas, may also be present but are smaller and more difficult to locate.

Spine: The primary structural framework upon which the fish's body is built; connects to the skull at the front of the fish and to the tail at the rear. The spine is made up of numerous vertebrae, which are hollow and house and protect the delicate spinal cord.

Spinal Cord: Connects the brain to the rest of the body and relays sensory information from the body to the brain, as well as instructions from the brain to the rest of the body.

Brain: The control centre of the fish, where both automatic functions (such as respiration) and higher behaviours ("Should I eat that critter with the spinning blades?") occur. All sensory information is processed here.

Lateral Line: One of the fish's primary sense organs; detects underwater vibrations and is capable of determining the direction of their source.

Swim (or Air) Bladder: A hollow, gas-filled balance organ that allows a fish to conserve energy by maintaining neutral buoyancy (suspending) in water. Fish caught from very deep water sometimes need to have air released from their swim bladder before they can be released and return to deep water, due to the difference in atmospheric pressure at the water's surface. Species of fish that do not possess a swim bladder sink to the bottom if they stop swimming.

Gills: Allow a fish to breathe underwater. These are very delicate structures and should not be touched by hand.

Kidney: Filters liquid waste materials from the blood; these wastes are then passed out of the body. The kidney is also extremely important in regulating water and salt concentrations within the fish's body, allowing certain fish species to exist in freshwater or saltwater, and in some cases (such as snook, tarpon, salmon, etc.) both.

Stomach And Intestines: Break down (digest) food and absorb nutrients. Fish such as bass that are piscivorous (eat other fish) have fairly short intestines because such food is easy to chemically break down and digest. Fish such as tilapia that are herbivorous (eat plants) require longer intestines because plant matter is usually tough and fibrous and more difficult to break down into usable components. A great deal about fish feeding habits can be determined by examining stomach contents.

Pyloric Caeca: This organ with finger like projections is located near the junction of the stomach and the intestines. Its function is not entirely understood, but it is known to secrete enzymes that aid in digestion, may function to absorb digested food, or do both.

Vent: The site of waste elimination from the fish's body.

Liver: This important organ has a number of functions. It assists in digestion by secreting enzymes that break down fats, and also serves as a storage area for fats and carbohydrates. The liver also is important in the destruction of old blood cells and in maintaining proper blood chemistry, as well as playing a role in nitrogen (waste) excretion.

Heart: Circulates blood throughout the body. Oxygen and digested nutrients are delivered to the cells of various organs through the blood, and the blood transports waste products from the cells to the kidneys and liver for elimination.

Gonads (Reproductive Organs): In adult female bass, the bright orange mass of eggs is unmistakable during the spawning season, but is still usually identifiable at other times of the year. The male organs, which produce milt for fertilizing the eggs, are much smaller and white but found in the same general location. The eggs (or roe) of certain fish are considered a delicacy, as in the case of caviar from sturgeon.

Muscles: Provide movement and locomotion. This is the part of the fish that is usually eaten, and composes the fillet of the fish.

## FISH DEFENSE STRATEGIES

Countershading-dorsally darkened and ventrally whitened, dark helps fish to blend in with the dark bottom when viewed from above whereas the white belly helps them to blend with the sky or clearer waters above when viewed from below.

Disruptive Coloration-another form of camouflage, colors and patterns (i.e.. presence of color stripes or bars) that break up the outline of a fish making it harder to see Eye Spot (false eye)-black spot located near base of the tail used to confuse predators

Thickened Scales-protective covering making their hard carpace relatively immune to predation

Spines-for defense and protection from predators, may be venomous Schooling-fish swimming in schools may have a greater chance to survive than if by themselves because an individual fish in a school may be harder to pick out by a predator

Almost three-forths of the world's surface is covered in water. This water is home to over 20,000 different species of fish. The earliest fossils of fish date back over 400 million years.

There are a wide variety of fish — from the goby which is less than one half an inch long, to the whale shark which can be over 60 feet long.

Most fish breathe through gills. Gills perform the gas exchange between the water and the fish's blood. They allow the fish to breathe oxygen in the water.

Fishes are vertebrates that have a skeleton made of either bone or cartilage. About 95% of fishes have skeletons made of bone. These bony fishes have a swim bladder, a gas-filled sac,that they can inflate or deflate allowing them to float in the water even when not swimming. Fishes with a cartilage skeleton tend to be heavier than water and sink. They must swim to keep afloat. Cartilaginous (cartilage) fish include the ray and the shark.

Most fish swim using a tail fin. Muscles in the tail fin move it from side to side, forcing water backward, and propeling the fish forward. Other fins help the fish change direction and stop. Pectoral fins on their side help them swim up and down. Dorsal and anal fins on the top and bottom keep the fish upright. Pelvic fins on the underside help steer left and right.

Many fish eat plants, while others such as the shark, eat other fish.

# 3

# FISH DISEASES

Just like human-beings and other animals, fish suffer from diseases and parasites. Fish defences against disease are specific and non-specific. Non-specific defences include skin and scales, as well as the mucus layer secreted by the epidermis that traps microorganisms and inhibits their growth. If pathogens breach these defences, fish can develop inflammatory responses that increase the flow of blood to infected areas and deliver white blood cells that attempt to destroy the pathogens.

Specific defences are specialised responses to particular pathogens recognised by the fish's body,that is immune responses. In recent years, vaccines have become widely used in aquaculture and ornamental fish, for example vaccines for furunculosis in farmed salmon and koi herpes virus in koi.

Some commercially important fish diseases are VHS, ich and whirling disease.

## DISEASE

All fish carry pathogens and parasites. Usually this is at some cost to the fish. If the cost is sufficiently high, then the impacts can be characterised as a disease. However disease in fish is not understood well. What is known about fish disease often relates to aquaria fish, and more recently, to farmed fish.

Disease is a prime agent affecting fish mortality, especially when fish are young. Fish can limit the impacts of pathogens and parasites with behavioural or biochemical means, and such fish have reproductive advantages. Interacting factors result in low grade infection becoming fatal diseases. In particular, things that causes stress, such as natural droughts or pollution or predators, can precipitate outbreak of disease.

Disease can also be particularly problematic when pathogens and parasites carried by introduced species affect native species. An introduced species may find invading easier if potential predators and competitors have been decimated by disease.

Pathogens which can cause fish diseases comprise:

- viral infections
- bacterial infections, such as Pseudomonas fluorescens leading to fin rot and fish dropsy
- fungal infections
- water mould infections, such as Saprolegnia sp.
- metazoan parasites, such as copepods
- unicellular parasites, such as Ichthyophthirius multifiliis

## Parasites

Parasites in fish are a natural occurrence and common. Parasites can provide information about host population ecology. In fisheries biology, for example, parasite communities can be used to distinguish distinct populations of the same fish species co-inhabiting a region. Additionally, parasites possess a variety of specialized traits and life-history strategies that enable them to colonize hosts. Understanding these aspects of parasite ecology, of interest in their own right, can illuminate parasite-avoidance strategies employed by hosts.

Usually parasites (and pathogens) need to avoid killing their hosts, since extinct hosts can mean extinct parasites. Evolutionary constraints may operate so parasites avoid killing their hosts, or the natural variability in host defensive strategies may suffice to keep host populations viable. Parasite infections can impair the courtship dance of male threespine sticklebacks. When that happens, the females reject them, suggesting a strong mechanism for the selection of parasite resistance."

However not all parasites want to keep their hosts alive, and there are parasites with multistage life cycles who go to some trouble to kill their host. For example, some tapeworms make some fish behave in such a way that a predatory bird can catch it. The predatory bird is the next host for the parasite in the next stage of its life cycle. Specifically, the tapeworm Schistocephalus solidus turns infected threespine stickleback white, and then makes them more buoyant so that they splash along at the surface of the water, becoming easy to see and easy to catch for a passing bird.

Other parasitic disorders, include Gyrodactylus salaris, Ichthyophthirius multifiliis, cryptocaryon, velvet disease, Brooklynella hostilis, Hole in the head, Glugea, Ceratomyxa shasta, Kudoa thyrsites, Tetracapsuloides bryosalmonae, Cymothoa exigua, leeches, nematode, flukes, Platyhelminthes, carp lice and salmon lice

1. Heavy breathing - General symptom
2. Stops eating - General symptom
3. Discoloration - Stress, or plistophora. Other diseases possible.
4. Bumps - Cancer
5. Swelling (especially around the stomach area) - Dropsy, overeating, constipation (more on this later)
6. Ragged fins - Fin rot or being nipped by other fish (more on this later)
7. Gasping (at the water surface) - Lack of oxygen, poor water quality, gill flukes
8. Clamped fins - General symptom
9. Inactive (for normally active fish) - General symptom
10. Whitish, stringy poop - Internal parasites
11. Cotton wool-like growth - Fungus
12. Specks (not from original pattern) - White spot or black spot disease, depending on colour
13. Cloudy eye - Bacterial infection or poor water quality
14. Sunken belly - Internal parasites or bacteria (be careful, may be Fish TB)

15. Swollen eye - Pop-eye
16. Holes in the fish - Hole in the head
17. Red gills (inflamed) - Ammonia poisoning or gill flukes
18. Red streaks - External bacterial infection
19. White threads hanging from fish - Parasitic worms
20. Erratic swimming - Possibly terminal stage plistophora
21. Pine-coning of the scales (when viewed from above) - Dropsy
22. Swimming upside down (except in synodontis cats) - Swim bladder disease or internal injuries
23. Struggling to maintain buoyancy - Swim bladder disease
24. Scratching against surfaces - Usually with the specks symptom, caused by external parasites
25. Swollen gills - Gill flukes
26. Dusty covering - Velvet

If any of those symptoms are present, you have cause to be worried. But calm down, take a deep breath and relax. Don't panic, we're here to help you. Write down the symptoms you have observed or the possible diseases, Remember, water quality is one of the most important things. Maintain good water quality, and it is exceeding unlikely for your fish to get any disease/parasite.

Good water quality is defined as (to me) - Ammonia 0, Nitrite 0, and Nitrate 40ppm and below

Now, false alarms

1. Swelling in stomach area - As mentioned earlier, may simply be overfeeding or constipation. Reducing the amount you feed and giving it boiled and shelled green peas will help out. Make sure it isn't another disease though
2. Ragged fins - It may be fin rot, which is caused by an infection or poor water quality. Or, it may simply have been nipped by a more aggressive fish. Watch it, and see if any other fishes are attacking it.
3. Floating at the surface, unable to swim down - May be caused by swim bladder disease, or simply that it swallowed some air bubbles while eating. If the latter, the problem should wear off within several days.
4. Exhibiting general symptoms of unhealthiness like lethargy, heavy breathing, stops eating, losing coloration, etc - If you have had that fish for a long time, a possibility you may want to consider is old age. Many smaller fishes only have a lifespan of 2-3 years, and will exhibit these symptoms at the ending of their lives.

If a fish is ailing, there are several steps that should be taken prior to using medication.

(1) Take a look at the ailing specimen, and take note of what is different from normal.

(2) Check to make sure that all environmental conditions are in order, because the number one cause of stress, ailing behavior, disease, and death in aquaria is

environmental problems. Test the pH, ammonia, nitrite, and nitrate levels, and if possible, the oxygen content. If these are in order, make sure that no other toxins such as paint fumes, pesticides, or cleaning agents could have entered the tank. If the answer is no again, make sure all accessories are operating properly (heater, air pumps, filter, etc.).

Try to see if another tank mate may have been on the aggressive. Run down the list of the fish's requirements to see that they are all in order. Remember that tank layout and dietary deficiencies also can affect the fish's health.

(3) If you find that one of these conditions is off, then take steps to fix it. If the water chemistry is the problem, make water changes until the problem is fixed. If everything checks out, and you still believe a disease is present you need to identify the symptoms of the disease and treat accordingly.

(4) Where possible place fish in hospital tank to treat and keep isolated from other fish for two weeks.

## ANCHOR WORMS (LERNAEA)

This is a parasitic copepod that attacks mainly freshwater fishes, but have also been found in marine habitats. These parasites can be seen with the naked eye. Females are 10-12mm, males are 0.8mm. Females are parasitic and embed themselves into the hosts flesh.

They are held in place using a 'holdfast', where they begin feeding on the fishes' flesh, organs and fluids. Reproductive potential is seen by the development of two egg sacs on the exterior/exposed part of the parasite, giving them that Y/T-shaped appearance. Young copepods are free swimmers, but soon attach themselves to the gills. Adults and eggs are introduced via live foods, infected water, or plants.

Symptoms: Fishes infected with these parasites can be seen 'flashing' on surfaces. This is the underside of the fishes as they attempt to rid themselves of these parasites. Other symptoms include localized redness, inflation of the body of the fish, breathing difficulties, lethargy, ulcers, dropsy, weight loss, loss of scales, gill and fin damage. Puncture wounds often introduce opportunistic, secondary infections.

Treatments: Infected individuals should be removed into a quarantine tank, as to prevent females from releasing their eggs into the main tank.

1. Salt has been recommended as the safest form of removal for effected fishes. 10-30 grams per litre.
2. Trichlorfon, Dipterex and Dylox (toxic to fishes and invertebrates - use with care)
3. Current treatments involves Dimilin

Anchor worms are actually crustaceans. The young are free swimming and borrow into the skin, go into the muscles and develop for several months before showing. They release eggs and die. The holes left behind are ugly and may become infected.

Treatment: The anchor worm is too deeply imbedded to safely remove. Treatment can best be done with a 10 to 30 minute bath in 10 mg per liter of potassium permanganate. Or treat the whole tank with 2 mg per liter, but this method is messy and dyes the water.

## Argulus

The fish louse (life cycle takes between 30 - 100 days dependant on the temperature.) Lice are one of the biggest threats to fish that you may come across. They are approx 5 to 10mm in size and can clearly be seen with the naked eye. The biggest danger from these parasites is from secondary infection such as; fungus, bacterial or internal problems. The best way to deal with this problem is to physically remove the lice with tweezers and treat the fish with an organophosphate based medication.

If you treat the fish three times over the life cycle of the parasite you will, almost always, eradicates the lice. Mortality rate. I had an infestation in one of my tanks and lost approx 60% due to a secondary systemic bacterial infection. Although I'm sure if I had caught it earlier this would've been lower

## Black Spot

Black Spot, or diplopstomiasis, often follows the addition of new aquarium fish. All fish are susceptible, especially the Silver Dollar and Piranha. It is fairly easy to diagnose and treat. Symptoms Small black spots on the body. Treatment Black spot is generally easy to cure. There are a number of commercially available treatments and preventatives.

## Boat

When a guppy shows a roughened appearance from the scales standing on end, and it appears to be bloated; it may have an intestinal infection which distends its intestines so greatly that it cannot swim below the surface without great exertion.

Some stay in an upright position.

Treatment: Try placing the bloated fish in a solution made of two tablespoonfuls of Epsom salts and two tablespoonfuls of Turk Island salt in a gallon of water. Leave it there for 4-6 hours. Then add another gallon of water and let the fish remain in this weaker solution for twelve hours before returning it to tank.

## Body Fungus

Body Fungus is just that, fungus. It will attack your fish's skin and cause ulcers or death in extreme cases. Large amounts of organic materials such as decaying food or fish waste can cause a breakout of fungus in your aquarium. Open wounds can develop fungus. The fungus feeds by excreting digestive enzymes which if it's on the fish, slowly digests parts of the fish. Though cured easily, if the fish doesn't get prompt treatment it will surely die. Secondary infections are also common.

Symptoms: White or grayish patches appear on the skin or gills that look like cotton or wool.

Treatment: Either purchase a treatment from your local fish/pet store specifically for fungus or buy an all round antibacterial treatment such as Melafix.

## Cataracts

Cataracts are fungal growths on the eyes. Treatment with any aquarium fungicide should work. Its probability increases with water rich in ammonia or nitrates.

Symptoms: White or gray material covering the eyes only.

Treatment: Special attention should be made to assure that ammonia and nitrite levels stay within accepted measures. Add anti fugal treatment. Raise the temperature and if fish are compatible with salt add it.

## Clamped fins

Clamped fins The fish clamps its fins close against the it's body. Experienced aquarists use this to quickly spot problems with their fish. Symptoms Like the name clamped fins.

Treatment: Water change (Like with all diseases) add melafix or some sort of appropriate medication. If possible add salt to compatible fish.

## Corneybacteriosis

Corneybacteria causes swelling in the head which will push the eyes outward. It is caused by overcrowding and water of poor quality, having an excess of ammonia and / or nitrites.

## Symptoms Bulging Eyes

Treatment: There are a few commercially available products which treat corneybacteriosis Penicillin and tetracycline are among them.

## Cotton Mouth

Mouth Fungus is so called because it looks like a fungus attack of the mouth. It is actually caused from the bacterium Chondrococcus columnaris. It shows up first as a gray or white line around the lips and later as short tufts sprouting from the mouth like fungus. The toxins produced and the inability to eat will be fatal unless treated at an early stage.

## Cotton Mouth

Symptoms: White cottony patches around the mouth and loss of appetite.

Treatment: There are commercial products available to treat cotton mouth.

## Cysts

This is an uncommon disease. Fish infected by cysts should be euthanatized before the cyst bursts.

Symptoms: Large balls appear on fish internally or externally. They look full of fluid and are relatively large.

## Dropsy

Dropsy is a bacterial infection that can attack the internal organs. This bacteria seems to only affect fish weakened by bad water conditions, stress, etc. Excess body fluid builds up making the fish appear bloated and the scales protrude outward from the body. This disease needs to be caught early to have a good chance of successfully treating the fish.

Symptoms: Bloating, protruding scales, exhaustion, isolation, loss of appetite.

Treatment: There are commercial treatments available from any fish/pet store, but it must be diagnosed early. Use Kanamycin or Tetracycline and raise the temperature to about 80 degrees/ 30 deg C. This helps remove the fluid note this is rarely curable

## Fin rot

Fin Rot is a bacteria that attacks the fins and in extreme cases, the body. This bacteria is usually the result of environmental stress such as poor water quality, bullying, etc. Part of treating this illness is early detection. If left untreated, secondary infections can occur. Fin or tail rot can also be a secondary infection caused by stress from other diseases.

There are many different ways that harmful bacteria can enter your aquarium, some fish at the pet store will look perfectly normal but harbor harmful bacteria that the fish has become immune to, and will release the bacteria through its feces in your tank. However, healthy fish will be able to fight off the bacteria with little trouble. If only a few fish infected, it is best to move them from the main tank, and place them in a bare hospital tank for treatment, but in most cases a hospital tank is not an accessible. Fish should never be given low dosages of medications over a long period of time; this could cause the bacteria to become immune to medications. Always follow the instructions on the bottle.

## Signs

Can be either bacterial (fins rot with red [blood] outline on edge) or fungal (white, cottony grows on edge of fin) The fins will have a ragged, uneven appearance that will eventually, if left untreated, disintegrate to the point where only a stub remains. Fin rot often begins with red streaks in the fins that are soon followed by abnormal lightening of the edge of the fin and then a fraying or rotting of the fin membrane and edges. This infection can lead to a complete rotting of the fin and can spread to the body and lead to death.

Symptoms: Rotting fins, loss of appetite, sluggish movements and laying on the tank bottom.

Transmission: The bacteria can be transmitted through the water from open ulcers, feces of infected fish, or through the consumption of infected, dead or dying fish in the tank. Predisposing factors: This disease is not highly contagious; however it is often associated with poorly kept or dirty tanks with poor water quality. Any dead fish should quickly be removed and disposed of. Diseased live fish should be removed to a treatment tank.

Treatment: The antibiotics kanamycin or erythromycins are sometimes successful. All infected fish should be handled with care to prevent transmission to humans through open wounds or cuts I had fin rot with angel and this fixed her up. For 10 US gallons put in half a cap full of melafix and half a cap full of pimafix for 7 days and do 25% water change on the seventh day.

Be sure to keep an air stone in there and don't have the temp to high. For one bacteria grows faster at higher temps, and these meds (which are natural...tea tree oil) will reduce O2 levels. I also cut down of feedings. I have experienced ammonia spikes in 10g tanks that were only mature for a month or two. I haven't had any problems with tanks more than 3 months mature. Something to keep in mind.

### Fish Tuberculosis

My goldfish tank got tuberculosis from a contaminated holiday food block. One of the fish was blind and not getting enough to eat so I'd been putting blocks in to feed him up a bit. Anyway, I was away for two days and when I came back ALL my goldfish had tb. I treated them with a 50% water change, aquarium salt, and melafix. Two died, two survived. I'm not sure if this was the best way to treat tb, but that's what I did... Flexibacter is a gliding bacteria. These bacteria are long, thin, and flexible. One end of the bacterial cell is attached to the fish, while the other end if free floating.

Symptoms: Looks like fuzzyness basically. Balls that look like mold. Treatment Appropriate medication (Maracyn and Maracyn II used together). Increased aeration, and decreased temperature.

### Fungus

Common disease that affects all kinds of tropical fish. Fungus mostly infects in poor water conditions in which there are unacceptable levels of ammonia or nitrites. Fin nippers will damage the fins of other fish making them more susceptible to fungal infections and external bacterial infections such as fin and tail rot.

Symptoms- A white cottony fluff will appear on fish. Can be around body or in mouth. Fin rot appears as white on ends of fins.

Treatment: Relatively easy. Add appropriate medication (Available for all fungal infections). If fish are salt tolerant add salt. Raise the temp to 30 deg C or 80deg F. You should treat the entire tank with fungicide to stop an outbreak. Determine the cause of the fungal disease and rectify it.

## FISH POX: MAINLY COLDWATER FISH

### Cause Viral infection

Symptoms- White, grey, or pink growths on skin and fins. Fish look like molten wax has been poured onto their bodies. The problem may advance and the skin might take on same color as surrounding tissue.

## Occurrence

Mainly coldwater fish, especially koi. The disease often appears, develops, subsides, and reappears later. Not very contagious and fish are rarely lost to it.

Treatment-"No reliable treatment." Raise temp by 5-10C or 9-18F might eliminate problem temporarily. This disease is unsightly, more so than dangerous, but avoid buying fish with this disorder. Bacteria are present everywhere on both land and in the water and are normally classified as either 'good' bacteria or 'bad' bacteria. The good bacteria in the aquarium consist of the ones that make up the biological filter.

These bacteria break down ammonia and nitrate so that it can be removed from the water. Without good bacteria most life would not be able to exist.The bad bacteria are often bacteria that are normally present but don't cause problems until the fish is injured, stressed or suffering from another disease.

These bad bacteria take advantage of the compromised animal's weakened immune system and reproduce extremely quickly, creating the resultantsicknesses and problems.The key to dealing with any bacterial infection is early recognition and treatment. Of course, prevention through careful introduction of new fish, plants and water, as well as maintaining a healthy, stress-free environment for your fish, is still the best course of action.Fish Pox: Mainly coldwater fish Cause Viral infection.

## Gill disease

Caused by Certain fungi, bacteria, protozoans and monogenetic flukes, and/ or poor water quality

## Symptoms

Rapid gill movement, swollen gills, and discolored gill filaments with excess mucus. Fish stop eating and lie motionless on bottom or gasp at the surface.

## Occurrence

Newly imported fish kept in bad water conditions or established tanks and ponds that aren't maintained properly. Poor filtration, overstocking, and infrequent water changes. Also might be caused by irritation of gills due to chlorinated water.

## Treatment

Improve the water conditions--25-50% water change. Administer antibacterial medications. If the problem persists, treat for parasites with an organophosphorous insecticide.

## Hole in the Head

Here are two different diseases referred to as hole in the head. One is caused by a flagellated protozoan, either hexamita or spironucleus and the other is environmental. The first is less common in the aquarium due to breeding.

The symptoms and sores caused by the disease differ. The end result can be fatal in both cases though. The first is so uncommon with the modern age of fish keeping that it is rarely found any more. It is caused by the specific pathogen a flagellatedprotozoan. The protozoan is found in every fish in their guts and other organs. It is kept in check in healthy fish through their immune system.

Once a fish is weakened or environmental changes favor the protozoan, the numbers of the protozoan in the fish increase thus resulting in hole in the head (HITH). The symptoms of HITH are the fish will lose its appetite suddenly, will excrete white slimy feces and will develop puss filled holes in the head area. Left untreated thiscan result in the fish's death. The other HITH is caused by environmental stress. This is the most common form of HITH and can be an indicator of poor water and filter maintenance.

This can be caused by an excess of fish waste in the water, but more than likely by the presence of heavy metals in the tank such as copper. More than likely in rocks that are gathered and not tested to be safe for the aquarium. the fish develops small open sores hat are not puss filled but rather like pits that can become enlarged as time goes on. This is caused by the epithelial cells dying.

This can continue till too much of the head is exposed to the open water and the fish dies. Unlike the other HITH the fish does carry on its normal appetite and produces normal looking feces. There is no pathogen responsible in this form of HITH.

While the first HITH can be cured with meds, the second form will only be aggravated by the meds. The first step is to find out the environmental cause of the stress either poor water quality or the contamination of heavy metals. You will need to change the water at least 75% every week to 10 days.

Remove any rocks that are not approved for aquarium use, i.e. out of your yard or found on the banks of rivers, lakes, etc... Another problem is the filtration of the water. If you choose a filtration system for a fish that is not full grown and do not regularly check that it is keeping up with the load of waste, this can cause excess waste in the water.

Aquarium filtration only slows the degradation of water quality it does not stop it. From looking at your water changing schedule, you probably are experiencing the latter of the two HITH. The Oscar is probably eating and pooping normally but he is growing. The water maintenance schedule was probably sufficient when he was younger, but now he is quite larger and you need to increase thefrequencies of changes.

Once you get rid of the HITH you should continue the water changes on a weekly basis of at least 20%. Also make sure that your filter system is large enough to handle the amount of waste that is being produced by your Oscar. Hole in head is a very serious disease that warrants prompt treatment. This is a common disease among larger Cichlids such as Severums and especially...Oscars. It appears as small holes on the face and around the eyes that may have a white material growing inside them. Hole in head, believe it or not, is an intestinal tract infection.

Symptoms: It appears as small holes on the face and around the eyes that may have a white material growing inside them. Loss of appetite and whitish feces are another indication that may appear before the development of the holes.

Treatment- A range of medication including Aquarium Pharmaceuticals-General Cure, Aquatronics- Hex-a-Mit, Aquatronics- Hex-A-Vital (especially for Discus, You can also treat with Bettafix and do a water change every week around 20%.

## INTERNAL BACTERIA

### Symptoms

a. In some cases fish will darken in colour, become listless, stop feeding and die.

b. In other cases, severe symptoms exist, ulcers, abdominal (dropsy) and eye swelling (pop eye), destruction of the central nervous system and then the fish dies.

Influencing Factors - Poor water quality will stress the fish and make them susceptible to infection. Also fin and body tissue damage will give the bacteria a site to start infecting.

My Experience -I recently lost one of my Mollies she was black so I did not notice her darken in colour, but she hid all the time and stopped eating, then died. At the point she died four other mollies and two platys started to hide and not eat. Also around this time started to notice that some of the fish had long stringy white feces (poo), it was 5 inches long in some cases. I did a 30% water change and added Interpet No 9 for internal bacteria, plus two spoons of salt.

Next day I added two more spoons of salt. On day four I put in the second dose required of Interpet No 9. Then nothing else and no water changes for the next 8 days. On the day four (after second dose) one of the mollies passed a lot of white gooey stuff and after that he started to come out at feeding time and occasionally take one or two nibbles. All fish were fine at the end of the 8 days, out from hiding and eating again. Some did however still have a bit of stringy poo. I was therefore advised to do a 25% water change and clean up and redose the medication without salt. This is mainly as a precaution because the disease can come back if it is not "completely" cleared up.

Reason it happened - I did not know at the time buy my pump housing was blocked my pump hardly running at all. My water parameters did not show there was a problem but the pump must have been getting slower and slower over a period of time and therefore the oxygen was getting lower and lower. I believe this is what caused the bad water conditions that kicked in the infection for my fish.

## What I have learnt

Fish commonly get ill because of poor water quality. I knew I had not introduced anything new, I did regular water changes and my tests were all good. How could it be poor water quality? It was because of the pump problem. I say don't give up looking there is always a reason, don't just treat the disease, find out why and learn from it.

## White Spot - Ichthyophthirius (Ich)

This is probably one of the most common fish diseases. Ich looks like literally white spots all over the fish. This is a Protozoan disease caused by Ichthyophthirius in fresh water aqauriums (Cryptocarion irritans in Marine). It is a parasite that attacks the eyes, fins, gills, skin and mouth interior Ich attacks the skin and gills of the infected fish. Ich is a parasitic infestation that is a very common pest among hobby aquarists.

It can be introduced by rapid changes in water chemistry such as temperature or ph (these weaken the fish's immune system leaving them susceptible to disease), poor water quality, or bullied fish. Ich is easily treated if caught in time. Once symptoms are observed, treatment must be swift. The life cycle of ich is about 4 days, if treated; it will have to be a minimum of 4 days. While the ich is in cyst form, it can't be killed by medication so you have to treat long enough for the cysts to hatch into its free swimming form.

Symptoms: White spots all over the fish. Fish will flash or scrape its self against aquarium décor, appearance of small white cysts or "sprinkles of salt" on the body or fins, loss of appetite, rapid gill fluctuation (breathing heavily), exhaustion, isolation, fish may hang out near the surface or water return. They will have a loss of appetite or stressed-rapid breathing. The parasite can only be affected by treatment during the free-swimming stages of the cycle.

Treatment: There are many different treatments available for ich, just go to any fish/pet store. A Copper Sulfate or Copper Formalin medicine should be able to kill most infections. There are many medicines all do the trick. If the fish are salt tolerant give your fish a salt bath. Many people believe that the addition of aquarium salt at the ratio of 1 tablespoon per 5 gallons of water, to be very effective along with the commercial treatment, and upping the temperature to 82F. Also raise the temperature a little if you can because this will speed up the ich lifecycle. White spot (ich) lives in the water so if your fish get it, you need to treat the whole tank and not just the affected fish.

The important thing about treating white spot is that you need to have patience. There is no quick cure. Once your fish have the spots there is nothing that can be done about them, what you're trying to do is kill off what is in the water. Treatment should continue for at least 4/5 days after the spots have cleared to make sure that the ich in the water is once again dormant.

## Neon Tetra Disease

Neon Disease is caused by a parasitic single called organism called Pleistophora hyphessobryconis. The cell, releases millions of spores in to the water during reproduction. The spores are consumed with food by the fish where they then enter the intestine. The spores grow and replace the muscle fibers weakening the fish. If an infected fish becomes wounded or dies, the spores can escape and infect a new host.

Symptoms: Pale white patches appear under the skin beneath the dorsal fin. The fish also becomes emaciated and may swim erratically.

Treatment: Commercial remedies are usually ineffective as by the time diagnosis is confirmed the fish is too weak to be saved.

## Pop Eye

Pop eye is an infection that can have a few causes such as bacteria, Poor water quality, injury, poor nutrition, tumors, and vitamin deficiently. The symptoms are quite clear.

Symptoms: Eye or eyes protrude from head, cloudy eyes from bacteria infection.

Treatment: Generally hard to treat, u may try a broad spectrum antibiotic.

## Saprolegnia

If on the fish's body, a white slimy, flattish patch appears which seems to have replaced a part of the fish's skin, it is probably saprolegnia.

Treatment: Isolate the affected fish and disinfect the net. (1) Add two drops of tincture of metaphen to each gallon of water in the tank to attempt to destroy the organism. Several other treatments of the sick fish are also recommended: (2) Salt treatment, (3) heat treatment, (4) Apply hydrogen peroxide to the spot while holding fish in a damp net. Let it isinfect for 15-30 seconds before returning fish to the aquarium. Be sure to re-disinfect the net. (5) Malachite green, (6) Methyline Blue, (7) Acriflavin.

## Septicemia

A fairly rare to common in all tropical and coldwater fish. It is caused by bad tank conditions and is very hard to cure if not impossible. This condition can follow on from skin infections such as fin rot or may occur independently as a result of dirty conditions. Bacteria enter the blood stream and circulate through the tissues causing inflammation and damage. Blood vessel and heart tissue damage result in leakage of fluids into the abdomen, producing dropsy. Inflamed blood vessels in the skin and at fin bases stand out.

Bacterial Hemorrhagic Septicemia can be diagnosed by red streaking of the fins and body and once this disease progresses, ulcerations and body sores start to appear. There is significant loss of appetite, lethargy, loss of motor functions in the later stages of the disease. Things To Look Out For Reddening at the bases of the fins Small hemorrhages around the eyes Very dull, listless behaviors Lack of appetite Treatment.

Seek vetinary guidance. The vet will prescribe antibiotics. I recommend using Oxytetracycline Hydrochloride Powder obtained from a vet to cure this disease. Check the aquarium for the cause and eliminate it. Diagnosis: The fish scrapes itself against objects; whitish-green threads hang out of the fish's skin with an inflamed area at the point of attachment.

## Skin & Gill Flukes

Flukes are a parasite that attack the gills and skin of infected fish. Small numbers of flukes are pretty much harmless, but in bad water conditions, over crowding, or aquaria with high organic content, they can multiply at a fast rate. According to the fish doctor the only reliable way to diagnose is with a skin scrape.

Symptoms: Fish will flash or scrape its self against aquarium décor, rapid gill movement, at a more advanced stage; the fish will isolate its self and lie at the bottom with fins clamped. The gills may turn red and swollen. Also cloudy skin due to excess mucus.

Treatment: Flukes are hard to treat and complete eradication is virtually impossible. It is said that for individuals affected with flukes, consecutive salt baths over 2-3 days can be useful. There are commercial treatments available

## Skin ulcers Signs

Ulcers usually show up as raw opened areas on the skin. They often have reddened edges and may be associated with other symptoms of systemic infection or disease.

Cause: Aeromonas, Pseudomonas, Mycobacterium and Vibrio.

Transmission: The bacteria are commonly found in the water and will invade a wound or skin injury on an otherwise healthy fish. Predisposing factors: Previously damaged skin, poor water conditions, and stress can all lead to an increase in ulcers. Treatment: Antibiotic baths coupled with anti-fungal baths containing phenoxyethanol are the most common treatments. Improving water quality, reducing stress, and decreasing the risk of injury are all very important in reducing the incidence of ulcers.

## Swimbladder

This is a disease that affects the fish's swim bladder. Swim bladder disease is sometimes caused by deformed bladder, parasite infestation, bacterial infections or constipation.

Symptoms: Fish swim side-ways or rest on bottom and can't go to surface. Loss of balance and/or abnormal swimming

Treatment: This is also a hard disease to treat; there is also a product around called Paragon II by Aquatronics® that supposedly aids in treatment of swim bladder disease. Raise the temperature; add appropriate medication for swim bladder. Note this is a disease that is rarely curable. Treat with Melafix every fortnight and salt everyday and a water hange every fortnight... not the same day as putting in the

Melafix. Feed Hikari Betta Pellets. All these treatment were done by me and most of them worked just fine. The treatment also worked with that fish that had the problem so please don't blame me if you try the same treatment to a different fish and it doesn't work.

## Tuberculosis

A fairly infectious bacterial disease, tuberculosis is becoming increasingly common. Affected specimens must be removed from the aquarium immediately so that other fishes aren't infected. A tubercular fish usually feeds normally, but loses weight as its internal organs become damaged. Some fish develop nodules under the skin which eventually ulcerate, in other nodules develop behind the eye, causing "pop-eye".

The bacterium that causes the disease prefers cooler emperatures than most bacteria that infect humans. However, fish tuberculosis can affect people, usually in the for of an infected nodule on the skin, but there is a small chance that it will cause a serious internal infection. Once diagnosed in one of your fish, strict hygienic precautions should be observed. A definite diagnosis is only possible by a post mortem.

Things To Look Out For

1. Appears dull in color
2. Weight loss
3. Folded fins
4. Ulcerous skin wounds

## Treatment

Seek vetinary advice. Affected fish should be removed and euthanized. Don't allow them to die in the tank as other inhabitants will eat them and become infected too. The tubercular fish's contacts should be treated: move them to a separate hospital tank, and disinfect the original aquarium. If other fish succumb, don't introduce any new specimens, euthanize all affected fish then clean, disinfect and re-stock the aquarium.

Tumor - There are 2 types of tumors. The Benign and Cancerous tumor. Symptoms: The benign tumor will have something that looks like an infection around it. A Cancerous tumor will keep growing and some scales around the area will protrude when it grows big enough.

Treatment: Sadly no cure. Either euthanize the fish recommended or let it live in agony for its remaining days.

## Velvet (Oodinium)

Velvet is a parasite that is often confused as ich. The difference is, velvet is smaller and infects predominately the body and looks like a fine powder rather than salt sprinkles. Velvet attacks the body of the infected fish and is very contagious. Velvet is most likely caused by stress, poor water quality or chilling (sudden changes in water temperature).

Velvet is a bit easier to cure than ich because the life cycle is shorter. However, this disease can prove fatal if left untreated. The parasite attaches to the skin and gills and large infestations can kill in a very short time

Symptoms: The parasites can usually be seen as a light dusting of gold spots on the flanks of the fish but can also appear as blue/grey fur like patches, if the parasite has entered the gill cavity then the gills will be flared and bright red in colour, fish will usually gasp at the surface. Fish will dart around and flash or scrape its self against aquarium décor, appearance of fine yellowish or white dust on body, fins clamped and/or rapid gill fluctuation (breathing heavily).

Treatment: Use a medication intended for Velvet/oodinium, most white spot/ ICH treatments will also work on velvet but check the label to be sure, if your fish allow for it add 1tsp per gallon of aquarium salts as well.

The parasite while in its free swimming form is photosynthetic and takes its energy from light, for the best results in curing the disease completely cover the tank to block off any light and leave the tank in darkness for seven days,any live plants should be removed at this time and treated in a separate container. Increasing the temperature to 82°F and application of a commercial chemical treatment is most effective.

Cause: A parasite similar to white spot/ICH velvet is present in most aquariums but will only infect fish and become a problem at times of stress, i.e. when water quality is poor or a fish has been newly introduced Fibrosis Fibrosis is most often found in marine or brackish water fish but can occasionally be found in tropical species as well.

Signs: There are two forms of fibrosis.

## Acute Form

Death may occur suddenly, before any signs are noticed. Symptoms may include increased respiration, loss of appetite, lethargy, skin hemorrhages, and death. Post mortem exams may reveal enlarged internal organs, but without a diagnostic test, this disease is difficult to distinguish from other bacterial infections. Chronic Form: Exophthalmoses, ulcers, and intestinal inflammation in fish that have died.

Transmission: Fish contract the bacteria Vibrato angular through open sores or feeding on dead fish that died from the disease.

Treatment: The best treatment includes the oral antibiotics chloramphenicol or furazolidone.

Summary These are just a few of the most common bacterial infections that can infect fish. After reviewing the list of infections, it is very clear that most bacterial infections are caused by a few similar situations, and that following a few basic recautions can prevent most of these infections.

The precautions include maintaining excellent water quality at all times, quickly removing any dead or diseased fish from the tank, treating all diseased fish, never introducing diseased or sick fish into your community tank, and preventing injuries from fighting or unsuitable habitat. If these basic guidelines are followed, bacterial infections will be a very rare occurrence in your tank.

## EUTHANASIA

For the aquarist the saddest job is that of euthanasia. This however is necessary in some cases, to avoid unnecessary suffering and pain to your fish. A lot has been written about this topic and there are many divided opinions. Acceptable Methods of Euthanasia. When considering euthanasia bear in mind that fish are capable of experiencing pain and stress and must therefore be destroyed humanely. There are a few acceptable methods and we will run through them one at a time.

1. Anesthetic- Immersion in an anesthetic solution such as Benzocaine or Tricaine Methane-Sulphonate is acceptable and may only be available via your veterinary practice. The effective dose will depend to some extent on the species of fish. An exposure to 300mg per litre of Tricaine Methanesulphonate is sufficient to cause the death to most aquarium fish. It is however advisable to leave the fish in the solution for at least an hour after death to prevent any possible recovery by the fish.

2. Decapitation- In many cases, especially with tropical fish this method is quick, clean and easy. With the aid of a scalpel, or a sharp knife or even a pair of very sharp scissors this job can be carried out with ease. Lay the fish on a flat surface, prevent it from escaping and with one quick motion, slice its head from its body. The body may still wriggle; this is acceptable, as it is just a nerve reaction. The brain will feel no pain. To prevent HYPOXIA, you must destroy the brain as soon as the fish has been decapitated. Simply holding the head and penetrating the brain with a sharp instrument does this. This may sound barbaric, but it is humane.

Concussion- This is a method that is preferred by a lot of aquarist. What you do is wrap the fish's body in a cloth, leaving the head exposed. Picking the body of the fish you then strike the head off the edge of something hard with a hefty blow.

This renders the fish unconscious and then again you must either decapitate it or destroy its brain to avoid any chance of recovery. Anglers' use a club, known as a priest to club big fish such as trout, to render it unconscious. I would not advise that you use such a device with tropical fish.

### Unacceptable methods of Euthanasia

1. Flushing a fish down the lavatory
2. Removing them from the water until death occurs
3. Plunging them in either boiling or freezing water
4. Freezing them to death, this method causes pain and immense stress. - Knocking them unconscious without destroying their brain.

### Safe Method of Disposal of the Carcass

You should now exercise some responsibility in the disposal of your fish. Do not throw it in the lavatory as the carcass may contaminate other fish in the waterways. Do not feed the carcass to your other fish or animals, it could spread disease. Remember the reason why you had to destroy it in the first place. The safest way is to wrap the carcass up in newspaper, place it in a polythene bag, securely seal it and dispose of it in the rubbish.

If you have a real fire, this is another effective method of disposal. Cremation Remember, Commercial fish medications are not regulated by any government agency. This means that some commercial medications have not been tested for safety or effectiveness. Thus some medications are completely useless and may be even harmful to aquarium inhabitants. Check out what you are buying, if it is compatible for the type of fish that you have. Try to find someone who has use that product before, or stick with well known brands that you can be sure of.

### There is something called Medication Poisoning

If medications are misused they can be harmful more than helpful. Medications can have adverse affects on many types of fish including catfish, tetras, mormyrids, loaches and other sensitive fish. Copper based medications have harmful affects on invertebrates, so always remove snails and crustaceans from the tank before treating it. Always be sure to read the label on medication to confirm that it is suitable for your fish. If medication appears to be harming your fish, make a partial water change and filter the water with activated carbon. There are a few things that you can do to prevent most of the diseases listed here which are:

1. Doing good water changes.
2. Keeping filters clean.
3. Not putting your hands in the tank a lot.
4. QT all new fish, for a min of a week.
5. Buy good healthy fish to start with, look at them at the pet store.
6. Give fish a hiding place. This can really cut down on the stress on the fish.
7. Live plants, since I have switched over to live plants I have had a lot less sick fish.
8. Don't add a lot of stuff (meds, stress coat, etc, etc) to the tank unless there is no other way around it, most of the short cut stuff is hard on fish.
9. Don't mix fish aggressive and peaceful fish should not be mixed, leads to fin nipping then in most times fin rot.
10. Don't treat fish till you're sure of what they got.
11. Watch temp swings in the tank.

## COMMON CAUSES OF FISH DISEASE

Aquarists often have a difficult time trying to figure out why fish die. The fact of the matter is there are many reasons why fish die. There are many known fish diseases that are fatal if not treated in its early stages. Fish will sometimes die without even being sick as well. This is just fish nature and is bound to happen.

Being able to spot the common fish diseases and knowing the causes will give you the upper hand when you need to combat a specific fish disease.

There are many reasons why fish end up getting diseases. While it is true that many diseases are caused by parasites and infections, we're going to discuss those that are not. New aquarists are more likely to hear about the diseases caused by parasites and infections. The diseases mentioned in this article are discussed less often. After reading the entire article, you'll understand the causes of common fish diseases. You'll also know how to properly monitor your aquarium for signs of fish disease.

## CARBON DIOXIDE POISONING

It is very common to see a news report detailing humans who have been poisoned by carbon dioxide. The symptoms and results are frightening. People usually don't realize that it is possible for fish to die of carbon dioxide poisoning. Aquarist need to take special precautions to ensure their fish don't fall victim to carbon dioxide.

There are a number of things to look for when trying to spot carbon dioxide poisoning in fish. Keep an eye on fish that seem lethargic and show signs of increased breathing. Be wary of carbon dioxide poisoning when your fish hang out near the water's surface. Carbon dioxide poisoning is caused when the water's aquarium lacks sufficient oxygen and has an abundance of carbon dioxide.

Aeration is extremely important during carbon dioxide poisoning treatment. Ensure that there is enough circulation in the aquarium. You can improve circulation using air stones or installing a water pump. Stop or cut back on plant fertilizer. Water changes should be performed more regularly to avoid carbon dioxide poisoning.

### Poor Water Quality

Poor water quality is one of the most common causes of fish disease. Pay close attention to your fish and take note of fish that are less active and stay around the water's surface gasping for air. Other signs of poor water quality include bad fish color and fish clamping their fins or closed fins.

Poor water quality is generally caused by insufficient filtration and water changes. Putting to many chemicals into your aquarium water can also decrease the water's quality drastically.

Everyone who owns an aquarium should have a water testing kit to test the quality of their water. To maintain good water quality, perform regular water changes. If ammonia, nitrite or nitrate levels are high, perform water changes on a daily basis. Continue and monitor water quality until they drop to safe levels. Proper circulation can really help improve the water's quality. Always maintain the proper pH level for the specific fish species in your aquarium.

## Metal Poisoning

Pale fish with strange behaviors might be a sign of metal poisoning. Luckily, this is one of the easiest causes of fish disease to prevent. Mental poisoning is caused when metal objects enter the aquarium's water. You should perform a complete water change should you suspect metal poisoning. Ensure that no metals enter your aquarium's water!

## Chemical Poisoning

Avoid using chemicals near your aquarium. This includes cleaning supplies, cologne, hairspray and insect killers. These chemicals can easily penetrate your aquarium and cause a whole lot of problems for your fish. When you absolutely have to use these chemicals near your aquarium, seal your aquarium shut using towels or plastic sheets.

Fish suffering from chemical poisoning will exhibit strangle behaviors such as refusing to eat, clamping their fins to their sides and laying or swimming on their sides. Pale colors and constant gasping for air are also signs of chemical poisoning. Treatment is simple. Perform a complete water change and also change all filter media.

## Improper Diet

Many fish diseases are caused by malnutrition. Fish suffering from malnutrition will become really lazy. Fading of colors on the fish is also a sign of a poor diet.

Fish diet is extremely important and plays a major role in determining how healthy your fish are. You should plan a well balanced diet for your fish as you would for yourself. Ensure that the aquatic animals receives the vitamins and minerals needed for the fish to thrive. Diets vary depending on your fish species. Normally this will involve commercial fish flakes. Deshelled peas should be added into the fish's diet as well.

## Overfeeding

Overfeeding could be classified as the opposite of malnutrition. Some fish species are more susceptible to overfeeding than others. Signs of overfeeding include lazy fat fish as well as constipation. Constipation can sometimes be remedied by feeding your fish deshelled peas.

Correcting overfed fish requires you to avoid feeding your fish for a few days. You'll also want to begin performing more regular water changes. The water changes will help remove uneaten foods from the aquarium which results in improvement in water quality.

Once overfeeding has been treated, you'll want to change the fish's diet to prevent additional overfeeding. Cut back on the quantity of food given to the fish each day.

## Frightened Fish

Fish can become scared by a number of different things. Fish that are stressed out or scared are more likely to harm themselves by crashing into aquarium decorations or even the aquarium glass. They'll also begin hiding and cowering as soon as the aquarium light is turned on.

This can be caused by a few different things including drastic lighting changes or people constantly bothering the fish by tapping on the aquarium glass. Other pets trying to get into the aquarium can also scare the fish.

Thankfully, relieving the frightened fish isn't all that difficult. The trick is to increase the lighting to the aquarium slowly before switching on the aquarium light. You can do this by turning the room's lights on or by opening the curtains closest to the aquarium.

# 4

# FISH DISEASES IN INDIA

In recent years intensive fish culture is being practiced in the brackishwater, coastal water and freshwater impoundments of India. Recent techniques of intensive and super intensive fish culture involve high rates of stocking and supplementary feeding which has substantially enhanced the incidence of diseases in fishes in our country.

In the first two decades of post-independence, lot of interest was paid to Inland fisheries and considerable amount of attention was paid to diseases in cultivable fishes in various states. Fishes suffer from a wide variety of diseases of infectious and non-infectious etiology in their natural environment. Number of literatures are available on fish mortalities due to bacteria, parasites, fungus, nutritional deficiencies, metabolic disorders and faulty environmental conditions like oxygen deficiency, high or very low water temperature, chemical pollutants, over crowding, mal-nutrition, algal blooms etc.

In India, traditional aquacullure has in recent years turned into a science based economic and commercial activity involving heavy inputs, and therefore diseases of all kinds are now known to occur on an increasingly larger scale. However, fish mortality is not the only criterion to evaluate the effect of fish disease. Even the morbidity, which leads to weight losses and poor growth in surviving fish, contributes substantial losses to the farmers. Since last decade there is an increased realisation of the necessity to avoid infectious agents to maintain healthy stocks to enhance production.

## Bacterial diseases

Innumerable diseases are caused in fishes due to bacterial pathogens and several of them are reported in Indian literature. They occur in nursery, rearing and grow out ponds causing serious concern to fish farmers. Some of them often wipe out the entire population offish. Some of the important bacterial pathogens are Aeromonas hydrophila, salmonicida, Pseudomonas fluorescens, Pseudomonas putrefaciens, Flexibacter columnar is, Edwardsiella tarda. Vibrio alginolyticus and V. parahaemolyticus which have been identified as most commonly encountered agents in fish diseases." Symptomatology, clinical pathology, microbiology and anatomic pathology have been carefully correlated to arrive at the diagnosis of the diseases caused by these organisms.

Hora and Pillay (1962) and Gopalakrishnan (1963) have reported fin-rot both in young and adult fishes. They have inferred that the disease is contagious and is capable of causing immense damage. However, none of them could attribute any particular pathogen for the disease. Recently, scientists at CI FA have reported that the fin and tail rot in young fish are due to a mixed infection of A. hydrophila and Pseudomonas fluorescens. Earlier workers suggested some control measures like bath treatment in 1:2000 copper sulphate for 1-2 minutes or painting of concentrated copper sulphate solution in the affected fish (Pal & Ghose, 1975).

Dropsy is another important fish disease in India where rohu, catla and mrigal are affected mostly in the late winter. Several workers have reported occurrence of the disease in composite fish culture ponds (Pal & Tripathi, 1978). The epidermis and body cavities get filled with fluid and scales protrude out from their pockets leading to severe anaemic condition. Pal and Tripathi (1978) although did not isolate the causative agent, could treat the affected fishes fully using terramycin in the feed. Kumar et al (1986) revealed a mixed infection of A. hydrophila and myxosporidian parasite in case of infectious dropsy in Catla catla. Histopathological changes in cases of dropsy have been described by theses workers. Mukherjee et al (1992) have attributed cases of dropsy of A. hydrophila in conjunction with malnutrition.

Another bacterial disease in the epidemic form has been studied by Gopalkrishnan (1961), and Kumaraiah (1977), the causative organism of which is identified as A. liquifaciens infecting the eyes of catla. The corneas of the eye gets vascularized leading to opacity and complete necrosis resulting in the death of the fish. Gopalakrishnan and Gupta (1960) have noted that optic nerves and brain are the primary sites of the infection.

Corneal opacity in silver carp {Hypophthalmichthys molitrix val.) due to Gram positive A bacterium, Staphylococcus aureus was reported by Shah and Tyagi (1986). Jhingran (1991) also could observe similar eye disease in epidemic form in Channa marulius and attributed it to the same type of organisms. Mukherjee and his co-workers (1992) have recorded a mass mortality in silver carps in a government farm and isolated Staphylococcus aureus from the affected eyes of diseased fish.

Chloromycetin bath @ 8-10 mg/litre of water has been found effective in controlling the disease at an early stage by some pioneer workers. Disinfecting the environment with KMn04 tit a dose of 0.1 ppm followed by liming @ 300 ppm check the disease as reported by Kumaraiah (1977).

Myxobacterial organisms have been attributed to a condition known as gill hyperplasia syndrome in common carp (Cyprinus carpio) as reported by Kumar et al (1986 d). The infectious nature of the disease has been proved by transmitting the disease in Asiatic major carp species through contact. These workers have conductedsome antibiotic therapy, but could not succeed in getting any significant result.

Ulcerative disease in catla has been reported by many workers where bilateral ulcerations of the opercula and the head are observed. In several cases the ulcerations have been reported to have penetrated the opercular bones and cranium. Gopalakrishnan(1963) and Karunasagar et al (1986) have investigated several such outbreaks. Kumar et al (1987) have studied this condition in composite fish culture ponds.

In most of thecases A. hydrophila could be isolated, although several other bacterial forms were also "present as secondary invaders. Mukherjee et al (1991) have reported the role of hydrophila in ulcerative disease offish and identified the biochemical differences among the various strains of these organisms. Nayak and Mukherjee (1994) made a detailed study of the biochemical properties of these organisms and drawn antibiograms on the basis of the antibiotic sensitivity tests.

In a recent study these workers have elucidated the role of A. hydrophila in dropsy, fin and tail rot and more elaborately in epizootic ulcerative syndrome. It is beyond doubt that in the ulcerative disease of fish these organisms might be playing an important role not only in necrotizing the muscle tissue but also damaging the internal organs viz. Kidney, liver and spleen.

Another disease of similar nature causing external lesions over the surface of the body has been described by Kumar et al (1986 C). The causative organism has been identified as Flexibacter columnaris. Topical application of potassium permanganate or short bath in 500 ppm of KMn04 has been found to be very effective in completely curing the disease.

A septicaemic disease called Edwardsiellosis has been reported by Kumar (1989) affecting brood fish population. Edwardsiella tarda has been isolated from the diseased fish showing anaemia, cutaneous lesions and gas filled abscesses in the muscle.

Although the mortality was not too high, but the morbidity has been reported to be extreme. Very recently, large-scale mortality in the spawn of Indian major carps has been reported by Mukherjee et al (1994) in an organised hatchery. E. tarda could be isolated from the disease affected spawn. The young ones showed anatomical deformities and greyish white discolouration of the body within 48 horus post-hatching and died within 72 hours. Although treatment with lodophor has been found to be effective, water quality improvement in the hatchery is the most essential component for keeping this disease away.

Epizootic Ulcerative Syndrome (BUS) is hitting the headline of the country since last six years and has become a matter of tremendous concern not only among the fishermen and fish farmers, but also among general public, entrepreneurs, administrators and planners. Initially the epizootic was mainly reported from the NorthEastern States of the country in the year 1988. Gradually, it has spread throughout the country.

At the moment it has engulfed the entire length and breadth of the country barring only few states like the Punjab, Jammu and Kashmir, Himachal Pradesh and Gujarat. Mukherjee (1994) has reported this not only in the freshwater fishes, but also brackishwater fishes in Orissa, Kerala, Kamataka and Goa. One common feature of the disease is that it initially affects the bottom dwelling species like murrels followed by catfishes and weed fishes.

Subsequently other fishes including Indian major carps are affected. Although the extent of economic loss has not been properly quantified, loss incurred by the fisheries sector due to this particular disease runs to several crores of rupees.

The lesions start as small grain to pea-sized haemorrhagic spots over the body, which ultimately turn into big ulcers of the size of a coin with greyish slimy central necrotic area surrounded by a zone of hyperemia. The disease affects the fish to such an extent that they start rotting while still alive and eventually die." A number of bacteria viz. Aeromunas hydrophila. A. punctata, Flavohaclerium Pseudomonas sp., E. tarda, V. parahae-molyticus and Streptococcus sp., have been isolated from the affected specimens (Kumar, 1989).

Histopathological studies revealed complete loss of epidermis in the ulcerative area of the skin where the dermis and hypodermis showed characteristic granulomatous changes (Dey, 1989), Gill tissue showed hyperplasia of the secondary lamellar epithelium, fusion of lamella and clubbing of gill tips. Haemopoietic tissue of the kidney showed proliferation of macrophages.

Besides bacteria, virus, fungus and parasites were also reported to be associated with epizootic ulcerative syndrome (Kumar el al., 1989; Dey, 1989).

Several methods have been tried or are being tried to control the disease. Many antibiotics, sulfonamides, chemicals, herbal preparations etc. have been advocated as preventive and curative measures. Lime was accepted widely among the fish farmers of the country until the formulation of a medicine by the scientists of CIFA. Marked improvement of the ulcerative condition is noticed within 7 days of application of the medicine and the ulcers are healed up within 10-14 days.

The medicine acts as a preventive also. Records reveal that there is no recurrence of the disease for at least eleven months in the ponds where the medicine is applied. The cost of application is only Rs.260/- (Approx. $ 8 US) for one Ha-m water body.

## Protozoan diseases

Myxosporidia are one of the most important groups of pathogen capaible of producing diseases in fish causing heavy loss in the juvenile. Several species of this pathogen have been reported from various parts of the country (Chaudhuri and Chakravarty 1970, Karuchandani, 1970; Senappa and Manohar, 1980; Sanaullah and Ahmad, 1980; Ahmad 1982; Dey et al 1988).

These workers have reported great problems and heavy losses in intensive fish culture and fish seed rearing due to emaciation and asphyxiation. Myxobolus cysts of varying sizes have been reported on At the gills and kidney of Catla calla. Larger cysts are located at the distal end of the gill filament where as smaller ones is seen at the proximal end. Infection damages the respiratory surface of the gill and excretory tubules of the kidney. Diagnosis of the disease can be made on the basis of gross appearance of the pin head sized greyish cysts and large number of myxosporidian spores under the microscope.

The other groups of protozoa infecting fish belong to ciliophora. The carp fry and fingerlings sometime carry heavy infection of the gills and skin with Trichodina indica.In heavy infection there is excessive mucus secretion on the gills which hampers respiration and the fish start surfacing. Some species of trichodina occur inside the urinary bladder. Under microscope they appear as circular transparent organisms with an '" internal disc-like structure. Fish affected with trichodiniasis eventually turn sluggish, lose weight and become moribund. There may be excessive mucus secretions and epidermal necrosis. A short bath in 2-3% salt solution for 5 minutes can control the infection effectively.

Whirling disease in Cirrhinus mrigala (Ham) observed by Maheswari (1986) has been attributed to a protozoan parasite Myxosoma cerebralis. The gross changes are characterized by deformed spinal column as reported in European and American salmonids suffering from whirling disease.

Ichthyophthiriasis or "white spot' disease has been reported by Gopalakrishnan (1963, 1964) causing large scale mortality in nursery and rearing ponds in different parts of the country. The fish disease unit of CIFA has also similar observations in carp nursery ponds. Presence of numerous pinhead shaped white spots on the skin, fins and gills are the common diagnostic features. On microscopic examination these spots look like blisters or nodules. Fingerlings of Labeo bala and C. mrigala have been experimentally infected with this parasite, but use of chemotherapy has not produced any encouraging result.

## Helminthic parasites

Fishes are Infected on their gills and skin by many species of monogenetic and digenetic trematoade and cestodes. The important monogenetic species are Gyrodactylus and Dactylogyrus. They remain attached to the skin or gills by the help of anchors which causes wounds. Their presence cause hypersecretion of mucous and haemorrhage from blood capillaries. These wound later get infected with pathogenic bacteria and fungi.

Alternate bath in 1:2000 acetic acid solution followed by bath in 2% sodium chloride solution has been found effective. Some workers have advocated dip treatment in 5% salt solution followed by bath in 1:5000 formalin.

The digenetic trematode generally infects the alimentary canal but are rarely harmful. However, in the metacercarial stage they are harmful as they get encysted in any part of the body, i.e. skin, muscle, abdominal cavity, liver, eyes and brain. In these cases tlsh serve as intermediate host while the final host is generally a tlsh-eating bird. In many.carp fingerlings the 'black spot disease' is due to metacercaria of diplostomatidae.

This was first reported by Ganapati and Rao (1962) in Calcutta. Catla and silver carp are the most susceptible species. Cestodes are also internal parasites in alimentary canal. Their larval forms are more harmful than the adults. They are found more in catfishes than in carps. The infected fish show softening of the muscle tissue probably due to accumulation of water.

In case of heavy infestation the abdomen bursts resulting in death of the fish. Bose and Sinha (1985) have described histopathological findings in cases of intestinal cestodes in Clarias hatrachus.

Acanthocephala have been recorded from major carps and cattlshes in India, but the life history or biology of the parasite has not been studied in detail. The parasite causes lesions in the intestinal wall by its proboscis.

Parasitic copepoda are external parasites infecting skin or gills or bucco pharyngeal cavity. Amongst them species of two genera Lernea and Ergasilus are reported from Indian carps. Lernea is reported to infect fingerlings and adults of major carps especially catla. The parasite penetrates the muscles and reaches the blood vessels as it feeds on the hosts blood. The tlsh thus become anaemic and weak.

Gopalakrishnan (1966) was of the opinion that larger number of these parasites may cause serious problems while a few may not cause any harm, Ergasilus sp. Infect the gill lamellae. They attach themselves to the gill by the help of their hook-like second antenna. Due to this there are wounds and haemorrhage in the gill tissue. There are no reports in India where mortality of tlsh has taken place due to Ergasilus infection.

## Crustacean parasites

Heavy mortality due to Argulosis or fish lice infestation has been reported in Indian major carps. Several species of Argulus have been reported to infect cultured fishes. These parasites inhabits the skin, fins and gills of the host and cause extensive pathological lesions in the skin showing circular haemorrhagic patches which become ulcerated. Mucous cells proliferate and copious mucous is produced. These lesions promote secondary infections like fungi and bacteria. Rohu has been observed to be the " most susceptible species among the Indian carps.

Histopathological changes in rohu naturally parasitized by Argulus spp. have been described by Dey (1989). Skin from the parasitized areas showed epithelial degeneration, oedema and hyperplasia of dense connective tissue Kidneys exhibited marked glomerular changes tubular degeneration and necrosis. In the liver, increased sinusoidal spaces and focal necrotic areas were also noticed.

## Fungal diseases

Gopalakrishnan (1966) observed Saprolegnia parasitica in the fry and fmgerlings of major carps. The disease in fish is characterized by a white to brown cotton like growth consisting of colonies of mycelium and filaments which appear as small to large patches on various parts of the body like fins, gills, mouth, eyes or muscle. Shrivastava and Shrivastava (1977) identified this infection in Channa punctata and opined that the infection starts due to netting injury and over-crowding or lesions caused by other diseases.

Jhingran (1991) reported that saprolegnia often infects the fertilized eggs. They attack the dead eggs first and then spread to the viable living eggs resulting in their spoilage. Shrivastava (1979) has reproted infection of Cirrhina mrigala with Aphanomyces pisci where the scales became black and later fell off. The hyphae of this species were not visible like those o\^ Saprolegnia. The author in the year 1982 reported other species of fungus, i.e. Achlya urion and Dictynchus anomalous.

Jhingran and Pullin (1988) have reported branchiomyces infection causing gill rot' in fish. They have opined that these pathogens invade the blood vessels of gills causing necrosis of the surrounding tissues also. Apart from these Aspergillus flavus, A. ochraceous and Fesiformis momiliforme have been reported from some of the EUS affected specimens (Mishra, 1992) at CIFA.

Many workers have recommended several treatment measures to counter fungal infection in fish which include dip treatment in 3% common salt solution or 1:2000 copper sulphate solution or 1:1000 potassium permanganate solution for 5-10 minutes.

A flush treatment for 1 hour in malachite green (0.1 ppm) is found to be quite effective in curbing fungal infection.

# ESTABLISHING DIAGNOSTIC WORK

## Introduction

The Government of India, assisted by the United Nations Development Programme and the Food and Agriculture Organization of the United Nations, are engaged in the intensification of Freshwater Fish Culture and Training Project (FI:DP/IND/75/031).

As part of the operation of the project, FAO assigned Dr N. Fijan as Fish Health Protection and Diseases Consultant from 12 October 1980 to 11 January 1981, with the following terms of reference:

1. to assist in developing the laboratory diagnostic methods for viral, bacterial, parasitic and other diseases and to initiate research programmes on fish health and diseases, and
2. to assist in developing the fish health monitoring system.

This report describes the work carried out by the consultant under the above terms of reference. In addition it contains proposals for long and short term approaches to laboratory diagnostic work and research on fish diseases, as well as on fish health monitoring at the Freshwater Aquaculture Research and Training Centre (FARTC), Dhauli.

## Background Information

Research at the Central Inland Fisheries Research Institute (CIFRI) on composite fish culture and implementation of results in field trials all over India have shown that intensive pond fish culture technology can secure very high yields (over 7 000 kg/ha) in this country.

From the worldwide experiences it is known that fish diseases represent a major threat to fish culture in general and especially to intensive biotechnologies. The personal contacts of the consultant with fish culture scientists revealed the presence of an experience or an opinion among some of them that, up to now, fish diseases do not seem to have a significant impact on composite fish culture in India, as managed by methods developed so far.

According to these opinions, losses in ponds can be avoided by already known methods in technology, including prevention of predation and of some parasitic invasions of skin and gills. On the other hand, some reports indicate that diseases do occur and have an impact on production in polyculture ponds. These reports as well as the limited laboratory diagnostic work and an experiment conducted at FARTC during the consultancy indicate that pond fish culture in India already stands at the same risks from diseases as the aquaculture in other countries.

Some information is available on fish diseases and their control in India. There are also some data on zoological aspects of fish parasites, on seasonal incidence of some ectoparasites and on gross symptoms of some diseases noted in composite fish culture. Generally, however, there is a lack of well documented research on most etiological groups of diseases and treatments for them. Bacterial and viral diseases were generally not a subject of up to date research by adequate laboratory methodologies.

Fish health monitoring is performed only to a limited extent at the field level. This situation indicates a serious lag in data on diseases and their prevention behind the research and practical results on fish production in ponds. In contrast to the situation in developed countries and in some developing countries, India does not yet have a diagnostic institution or laboratory for fish diseases. The Project IND/75/031 has therefore included work on fish diseases as one of its main objectives.

The FARTC at Dhauli is the first location in India that will have facilities and personnel for research and diagnostic work on all major groups of fish diseases and work on health monitoring in aquaculture. In order to fulfil its future role in the development of this field in India and in countries of this part of world that do not yet have such facilities, the initial planning of work on fish diseases is of utmost importance and can have far reaching consequences. The consultant has therefore included in this report general recommendations on organization of the work and on possible future plans.

## SUMMARIZED DESCRIPTION OF WORK

The consultant stayed at FARTC, Dhauli, from 16 October 1980 to 6 January 1981. The stay was prolonged for a dew days upon the request of the project director in order to complete some experiments. A four day break in above stay was made from 27 to 30 December for a visit to Barrackpore on the request of the Director of CIFRI.

There the consultant informed the director about the progress in his work at FARTC. He also went to two research stations of CIFRI to investigate two fish disease problems and to advise the research workers about possible research approaches for solving the etiology of these diseases.

Four young FARTC scientists R.K. Dey, D. Kumar, B.K. Mishra, K. Suresh) were assigned for work in this field. In addition, two CIFRI scientists (R.N. Pal and D.M. Swami) worked with the group for about one month each.

# TRAINING

The work of FARTC staff scientists could be started only after adequate training. The programme of training was adjusted to the fact that none of four assigned FARTC scientists had sufficient previous knowledge of fish diseases, fish health monitoring and protection. Special emphasis was placed on topics that are not fully covered in existing English or other literature, and on practical training.

## Theoretical Background

The following topics were covered during the training:

- General pathology, main types of pathological processes in fish and their symptoms;
- Characteristics of etiological groups of fish diseases;
- Epizootiology of fish diseases;
- Prevention of diseases in aquaculture;
- Methods of preventive and curative application of drugs to fish;
- Incorporation of health protection measures into pond culture technology;
- Selection, packing and transport of fish samples for examination in the laboratory;
- Diagnostic methods and procedures for examination of fish (gross examination, determining the condition of fish dissection, hematocrit, bacteriology, virology, parasitology).

## Practical Training

This was carried out on the following topics:

- Sampling of fish from ponds for examination;
- Laboratory examination of fish for gross lesions by dissection;
- Taking of material from fish for hematological, parasitological, virological and histopathological examinations;
- Microscopic examination of local preparations from external and internal organs for pathological changes, parasites and bacteria;
- Bacteriological examination of fish (preparation of smears, stains and media, inoculation and examination of dishes with media, basic methods for identification of fish pathogenic bacteria, maintenance of isolates);
- Primary cell cultures and cell lines for fish virology (preparation of glassware, media and other materials, preparation of cell cultures and their use for diagnostic work);

- Intraperitoneal administration of drugs to fish;
- Designing and planning of pond experiments on diseases, as well as their prevention and treatment; conducting of experiments and interpreting the results; an experiment was designed, planned, carried out and completed in 12 experimental ponds; a second experiment was started;
- Keeping records on laboratory work and findings.

### Organization of the Seminar on Fish Health Protection

Upon completion of the training by the consultant on general aspects of fish diseases, seven scientists gave a seminar for participants in the ongoing course at the KVK and Trainers' Training Centre of CIFRI. The purpose of this seminar was twofold: (i) to give an opportunity to the scientists that were trained by the consultant to confirm and deepen their knowledge, and (ii) to provide trainees at the Centre with basic knowledge in this field and to initiate cooperation in fish health monitoring between field workers and the FARTC researchers.

### Other Activities

The consultant gave three lectures on fish diseases and health protection. In all three instances fruitful discussions developed, showing great interest in this field and its importance. Lectures were held for (i) Director and Extension Officers of the Department of Fisheries, Government of Orissa, (ii) staff members and students of the Veterinary School at the Orissa University of Agriculture and Technology, and (iii) scientists of CIFRI at Barrackpore. The lecture under (i) was of special importance for the development of cooperation in work on fish diseases and health protection between FARTC and the Department of Fisheries.

## AID IN ESTABLISHING ADEQUATE CONDITIONS FOR RESEARCH

### Utilization of Space

Four laboratories and one storage room were allocated by the project director for work on fish diseases and health protection. The first laboratory will serve as the room for dissection and general examination of incoming fish samples, as well as for work on diseases caused by parasites. The second laboratory will accommodate work on histopathology and on bacterial diseases.

The third room will be the laboratory for cell culture and virology. The fourth room will serve as the wet laboratory for experiments with fish in aquariums and provide space for washing and sterilization of glassware. Depending on developments in the research and on practical factors, the space may have to be enlarged in the future.

The consultant made recommendations for setting up the sterile chamber for cell culture and virology work, needed adaptations of furniture and suggested plans for setting up the wet laboratory.

## Installation of Equipment

Together with the local staff, the consultant installed the equipment that came to FARTC from the Cuttack Station of CIFRI. When this equipment arrived, it was found that it did not function properly and a claim was sent to the supplier.

## Other Activities

In order to secure information about availability of commonly used medicines, the consultant requested and obtained information and literature from several pharmaceutical companies in India. Collection of such information will be continued by staff to have on hand broad information about locally available drugs for research and field use and their prices.

To facilitate the diagnostic and research work, the consultant prepared up to date flow charts for isolation and identification of fish pathogenic viruses and bacteria. The charts were to be prepared as wall posters.

The consultant also discussed with and advised the FARTC staff and the potential producers about the design of experimental aquaria, troughs and tanks that will be procured by FARTC for research purposes.

Cooperation of scientists assigned to work with the consultant and other researchers and research groups was initiated. On one hand, diagnostic information and preventive treatment recommendations were given for experimental fish. On the other hand, advice and cooperation was utilized or planned with economists, nutritionists, water chemists and statisticians.

# RESULTS OBTAINED DURING PRACTICAL TRAINING

During the consultant's stay in Dhauli, 101 fish were dissected and their organs examined microscopically. Material for bacteriological examination was taken from 28 fish, for parasitological examination from 10 specimens and for histopathological examination from 35 fish. In addition to pathoanatomical diagnoses, tentative bacteriological diagnosis was set up for 20 fish, final histopathological diagnosis in 4 fish and parasitological diagnosis in 82 fish.

Materials under processing and isolated strains of bacteria will be used for setting up final diagnoses and for research purposes.Examined according to species were 66 rohu, 10 catla, 7 mrigal, 3 common carp and 15 silver carp.

Of the above fish, 66 originated from ponds at FARTC, 27 from nearby village ponds, 5 from the Trainers' Training Centre ponds, and 3 from the Cuttack hatchery. Another break-down from the records shows that 76 of these fish were examined directly after netting from ponds. Two of them had malformations, while the others did not show external signs of disease. There were 25 fish brought for examination from laboratory experiments set up by groups of scientists working on nutrition and on genetics or from the pond experiment set by the fish disease group.

The most striking finding was that 100 percent of the rohu examined suffered from severe pathological changes in their kidneys, detectable only by microscopic examination. Changes seemed to be more pronounced and rather severe in two year fish. They are sufocate to chronic. Findings were confirmed by histopathological examination. Fish had necrotic and degenerative changes in most of tubules (nephrosis) and focal accumulation of a yellowfish material, which was sometimes surrounded by a weak chronic connective tissue encapsulation.

The yellowfish micronodules were less abundantly present in liver and in one case very abundantly in ovaries. In most of these fishes, myxosporidian specks were detected in kidneys. It is difficult to give an opinion now whether there is a causal or any relationship between the myxosporidian parasite, the nodules and the nephrosis. Since such a 100 percent rate of heavy kidney damage, as well as the occasional damage of other organs, definitely must have an influence on the condition, growth rate and food utilization of the affected rohu, this problem requires thorough further investigation.

The second important finding in the laboratory was that, after netting and handling (material available so far was limited only to fish in experiments), rohu and common carp suffer from symptoms and mortality typical of the columnaris disease. A Flexibacter columnaris-like myxobacterium was isolated in pure culture from kidneys of several affected specimens and was abundantly present in skin lesions.

This problem is of known significance in warmwater pond culture of North America and Europe. The disease can affect almost all fishes but there are some variations in species susceptibility. In some species it occurs almost regularly after netting and stocking. It also develops spontaneously in open waters and ponds, often affecting only gills. The findings here indicate that, in all cultivated species, prevention of this disease may have practical significance for reducing losses after netting, handling and stocking operations. Research in this field is urgently needed.

A possibly significant problem was detected by diagnosing enteritis in mrigal and the finding of the intestinal protozoan parasite from the germs opalina, which has not been reported in fishes from temperate zones. Further work on fish health monitoring should indicate the incidence and importance of this condition and the role of this parasite in it.

Presence of yellow clusters of cyst-like units in the intestinal content of some fish, which could eventually be spores of evecidian parasites, also requires further monitoring and investigation.

It can be concluded that health monitoring in ponds and diagnostic work on fish disease cases in India can provide more than ample material for research on fish diseases and their prevention, thus opening possibilities for incorporation of better health protection measures in aquaculture for higher and more economic yields.

## Fish Cell Cultures

These are the essential prerequisite for diagnostic and research work on viral fish diseases: preparing and maintaining of cell cultures requires a special laboratory environment, washing and preparation of glassware, media and other supplies, as well as the application of a technique that is sophisticated for the newcomer to this field.

At the end of the consultant's assignment, only improvized conditions for this work could be secured. One attempt to prepare primary cell cultures from rohu could be made and partial success had been achieved. The continuation of this research work and early securing of prerequisites for proper methodology are essential for progress in research and diagnosis of viral diseases.

# POND EXPERIMENTS

During the consultant's assignment one pond experiment was completed and the second one started.

In addition to training, the aim of the first experiment was to determine the influence of poor fish handling on losses and the possibility of preventing them by use of antibiotics.

The experiment was carried out in 12 ponds that were randomly used for four experimental treatments of rohu (650 g). Two stocking densities were used. Six ponds (two densities) were stocked with fish that received penicillin plus streptomycin intraperitoneally, while fish for the other six ponds received sham-injections of distilled water. All fish were treated equally otherwise. Their allocation to groups was also randomized.

The experiment lasted 12 days. Mortalities occurred 2 to 5 days after stocking and then ceased. All sick and dead fish exhibited symptoms of columnaris disease, while some also had saprolegnia infection of the skin. Flexibacter columnaris was isolated from skin lesions and kidneys of sick and dead fish.

Mortality in fish that received the antibiotics by injection was 1.75 percent and the mortality in non-antibiotic-treated group was 27.18 percent. The stocking density did not affect the mortality rate.

The above experiment indicates that an outbreak of columnaris disease can be triggered in rohu by handling. This is a known fact in fish culture of warmwater species in Europe and North America. The results obtained suggest that injection of suitable antibiotics can decrease losses after handling of broodstock during spawning operations, and that further experiments are needed to assess losses during normal stocking of ponds due to handling, as well as on prevention of such losses by bath or oral treatment of fish prior to stocking.

The second experiment was planned to find out if treatment with a nitraofuran compound can reduce the myxosporidiosis and kidney damage in rohu.

## PROPOSALS FOR ORGANIZATION OF WORK

The health of cultivated fish can be adversely affected by numerous biotic and abiotic factors. The number of these factors in fish medicine is even larger than in human or veterinary medicine. Research on fish diseases and fish health protection encompasses, therefore, a constantly increasing number of narrow research fields.

As in other research areas, development of sciences and their methodologies now leaves very little space for encyclopaedic oriented type of investigations. Bearing in mind that research on fish diseases has first to gain more ground in India and that practically oriented work must have priority, only a few not quite so narrow fields can be covered at present.

All young scientists starting to work on ichtyopathology and health protection at FARTC should develop knowledge and research techniques on general problems and prevention of diseases. In addition to this, each of them should concentrate on work in one field, according to specific research techniques.

In the first period of research development, general examination techniques for fish, basic bacteriological, parasitological, virological and histopathological techniques, as well as experimentation on prevention and treatments of diseases, should be mastered and used in diagnostic and research work.

Scientists assigned to these fields should work as a homogeneous team, supplementing each other's activity. This is essential for avoiding situations occurring sometimes in some countries with narrow specialists, where the same fish mortality may be attributed to quite different diseases by different researchers, thus confusing the fish farmer or the institution seeking advice for prevention or treatment of diseases.

For the development of working fields it is essential that a significant part of the activity be devoted to diagnostic examination of fish from as many ponds as possible from a large territory. Such activity can be secured through research programmes on health monitoring and taking inventory of disease problems in pond culture. This work is an essential prerequisite for proper selection of research priorities in ichthyopathology, as well as in fields of prevention and treatment. On the basis of diagnostic findings, research can be oriented towards the most urgent needs of aquacultural practice.

Within the diagnostic work, every effort should be made to ensure that fast and reliable diagnostics, together with proper recommendations for disease prevention and treatment reach the fish farmers or owners of the fish as soon as possible. Proper organization of work and development of appropriate methodology are needed for such performance.

## UNIT FOR ICHTHYOPATHOLOGY AND FISH HEALTH PROTECTION

For reasons mentioned in 4.1 above and in order to comply with proposed plans on research approaches and responsibilities at FARTC as the Regional Lead Centre for the project RAS/76/003, the consultant suggests that the four scientists assigned to the work on fish diseases start to function as the Unit for Ichthyopathology and Fish Health Protection. Training and utilization of assigned space has been carried out with this proposal in mind.

## Research Fields for Future Planning

Full coverage of fish health problems will be of growing importance for development of aquaculture in India. It is useful therefore to foresee all main research fields which will have to be developed either immediately or at a later stage. These fields may be listed as follows:

i. Fish health monitoring;
ii. Viral diseases;
iii. Bacterial diseases;
iv. Mycoses and algal diseases;
v. Protozoan and methazoan parasitic diseases;
vi. Environmentally caused diseases and stress;
vii. Nutritional diseases;
viii. Intrinsic and hereditary diseases, neoplasms and malformations;
ix. Toxicology;
x. Histopathology;
xi. Prevention and treatment of diseases and their incorporation into technology;
xii. Immunoprophylaxis.

The list may be useful for long-term planning of personnel, space, facilities, training and equipment.

Programmes covering some aspects of fields listed as i, ii, v, x and xi should be initiated immediately, as suggested in subsequent paragraphs of this section and in section 5.

## General Diagnostic Work

This basic work should be performed by all four scientists of the unit and should consist of activities described below.

Selecting, packing and transporting fish samples from ponds to laboratory, checking main water quality parameters on the spot and taking water samples. Accepting samples of cultivated carps and taking case histories from extension service personnel, state fisheries department, farmers or staff members of FARTC. Recording arrivals of samples at the laboratory. Keeping records on case histories, findings, diagnosis and recommendations. Preparing annual report on diagnosed cases.

Storage of live or dead fish samples in aquarium, refrigerator or freezer if necessary.

Clinical examination, dissection, macroscopic and microscopic examination of effected and suspect organs and tissues; taking of material for hematological, histopathological, parasitological, bacteriological, mycological, virological and other examinations, as well as for biological experiments (transmission of disease, if necessary for diagnosis).

After pathoanatomical diagnosis and collecting data on results of other examinations, setting up of final diagnosis. Quick reporting of final diagnosis and recommendations for prevention and treatment to the owner of examined fish.

Surveys of pond fish populations by sampling for presence of pathogens and diseases throughout the production cycle, with the purpose to identify the health status and eventual needs for preventive treatments. In this way detect specific season - and age - dependent health problems of fish in mono-and polyculture.

Cooperation with scientists in the centre in their work on various aspects of aqua-culture by checking fish health in experiments and recommending prophylactic and therapeutic treatments and advising on health protection measures in planning and conducting research.

## Virology

Registering samples for virological examination, recording of methods applied and of findings, release of virological diagnosis for setting up final diagnosis. Making recommendations for prevention. Preparing of annual report on diagnostic work and research.

Preparation of media and solutions for cell culture work. Subcultivation, maintenance and cryopreservation of existing suitable American and European fish cell lines. Preparation of primary cell cultures from indigenous and other cultivated fish species. Development of fish cell lines from indigenous and other important species for this region.

Extraction of virus and decontamination of inocula for cell cultures or experimental fish. Inoculation of cell cultures, incubation and detection of virus presence. Presumptive and confirmative identification of virus, using morphological, cultural, biochemical and serological characterization. Inoculation of fish, observation of results.

Research on and development of advanced and improved fast virological methods. Use of rabbits for preparing of antisera.

Study and description of viral disease entities by using epizootiological, pathoanatomical, histopathological, virological and serological methods, and biological experiments in aquariums, tanks and ponds.

Surveys of pond fish populations for presence of viral pathogens and diseases in various phases of aquaculture technology. Detection of virus carriers among broodstock and fingerlings.

## Bacteriology

Registration of samples for bacteriological or mycological examination; recording of method used and findings; quick release of presumptive (if possible also confirmative) microbiological diagnosis for use of setting up final diagnosis, and recommendations for treatment and prevention. Keeping records on recommendations and experiments. Preparing annual report on diagnostic work and research. Preparation of methodology sheets.

Preparation of media and solutions for isolation of various pathogens, determination of their biochemical characteristics essential for classification and diagnosis and for maintenance of isolates. Preparation of stains and staining of smears.

Plating of materials and isolation of pure cultures. Examination of characteristics of isolates (morphology, culture, biochemical and staining characteristics), performing of tests for serological identification.

Research on development of advanced and improved bacteriological diagnostic methods, including immunological and immunofluorescent ones.

Study and description of bacterial and fungal disease entities by epizootiological, pathohistomical, histopathological, microbiological and serological methods, and biological experiments in aquariums, tanks and ponds. Determination of pathogenicity and virulence of isolates.

Surveys of pond fish populations for presence of bacterial pathogens and diseases in various phases of aquaculture technology.

Study of normal bacterial flora of species in aquaculture (skin, gills, intestines, blood) and its role in diseases.

## Parasitology

Registration of samples for parasitological examination, recording of findings, quick release of parasitological diagnosis for use in setting up final diagnosis and recommendations for treatments. Keeping records on findings, recommendations and experiments. Preparing annual report on diagnostic work and research. Preparation of methodology sheets (procedures).

Application of various parasitological methods (flotation, digestion, concentration) for detecting protozoan parasites; fixation, staining and taxonomic examination of parasites.

Study of life cycles and ecology of parasites in pond culture, as well as epizootiology of diseases caused by them. Qualitative and quantitative surveys of parasitofauna in various stages of aquacultural technology.

Defining disease entities caused by parasites (epizootiology, histophathology, etc.). Aquarium and pond experiments on these diseases.

## Histopathology

Registering samples taken for histopathological examination, recording of methods used and of findings, quick release of histopathological diagnosis for setting up final diagnosis. Preparing annual report on diagnostic results and research.

Preparing of fixatives, histological stains and solutions, shaping up tissues for slicing, slicing by freezing microtome, embedding of samples in paraffin, preparation of paraffin sections, staining, mounting, microscopic examination, case studies and diagnostic work, preparing photomicrographs of sections.

Preparation of methodology sheets (procedures). Study of normal histology of various fish species of importance for aquaculture. Study and description of histopathology in individual disease entities caused by various etiological agents (including nutritional and toxic ones).

Cooperation with fish nutritionists and other research groups by providing histopathological expertise on status of organs in various experiments.

### Treatments

Preparation of lists of drugs for prevention and treatment of fish diseases with indication of source, price and cost of treatment.

Planning and conducting tests on suitability of various drugs for prevention and treatment of diseases in aquariums, tanks and ponds. Keeping records on tests, evaluation of results.

Preparing and giving recommendations on prevention and treatment of diseases in field situations. Keeping records on such treatments and evaluation of results.

Improvement of techniques for treatment of fish by bath, by food and by mass injections.

### Collection of Specimens, Pathogens and Slides

Accumulation of documentation for research reports, papers, training purposes and extension work should be a constant activity of scientists in the unit. This, as well as collection, maintenance and preservation of pathogens is also necessary for national and international exchange with fish disease laboratories. A reference laboratory for fish diseases does not exist in this part of the world and the FARTC may assume this role through the unit.

Setting up a museum of wet preparations of sick fish, collection of pictures of gross lesions and microscopic in fish, maintenance of fish cell cultures, storage of virus isolates, bacterial pathogens, specimens of parasites, histopathological slides, etc., should be a part of efforts to upgrade the quality of their own work.

## AID IN SETTING UP RESEARCH PROJECTS

The objectives are to develop methods of assessing the health status and the occurrence of subclinical and clinical diseases from the stocking to the fishing out of ponds.

### Practical Utility

By determining health status and specific disease problems, possibilities will be opened for application of prevention and treatments, thus increasing chances for higher yields. Results of monitoring will also indicate high priority research problems in the field of fish diseases.

## Technical Programme

The monitoring should be carried out in three production ponds with high stocking density managed for intensive culture.

Monitoring at three levels will be carried out and compared:

(i) Fish culturist level (farm workers). Daily observation of ponds for fish in distress or dead fish; daily observation of food intake; immediate notification of unit about unusual observations or drop in food uptake.

(ii) Aquaculture specialist level. Extension worker should observe the pond situation at least once every ten days and check fish for growth rate and general appearance by netting at one month intervals. At stocking and at fishing out, the number of fish should be precisely determined to calculate exactly the percentage of loss for correlating this with findings about fish health. Notification of unit when unusual findings are present.

(iii) Fish disease specialist level. To determine the health status and detect disease or parasite incidence at a 10 percent or higher level, a sample of 30 fingerlings of each species will be examined in the laboratory before stocking and 30 specimens of each species will be examined in the laboratory at bimonthy intervals and at fishing out.

In the unit, the following examinations will be undertaken: total length, weight and condition (Fulton's formula) of each specimen; determination of hematocrit value; dissection, macroscopic and microscopic examination of organs for detecting abnormal appearance of organs and presence of parasites; bacteriological examination of lesions (if they exist) and of kidneys of five fish from each species sample; histopathological sampling of abnormal looking organs, and sampling for virology if symptons suspective for a viral disease exist.

# 5

# VARIOUS TYPES OF FISH DISEASES

## VIRAL DISEASES

### Lymphocystis Disease

A. Iridovirus
B. Observed in most freshwater and saltwater species.
C. Clinically, fish are presented with variably sized white to yellow cauliflower-like growths on the skin, fins, and occasional gills. Occasionally, this virus may go systemic with white nodules on the mesentery and peritoneum.
D. Histopathology: Fibroblast undergoes cytomegaly with many basophilic cytoplasmic inclusion bodies and a thick outer hyalin capsule. The inflammatory response is variable but is usually a chronic lymphocytic inflammatory infiltrate.
E. The disease gains entry through epidermal abrasions. The virus infects dermal fibroblasts.
F. The disease is self limiting and refractory to treatment. Nodules may last several months and cause infected fish to be susceptible to secondary bacterial infections. Reinfection can occur.

## HERPESVIRUS SALMONIS (HERPESVIRUS DISEASE OF SALMONIDS)

A. Herpesvirus
B. Disease is observed primarily in the fry of rainbow trout.
C. Clinically the fish are lethargic with prominent gill pallor. Mucoid fecal casts are commonly observed trailing from vent.
D. Lesions:
    1. Exophthalmus and ascites
    2. Low hematocrit and numerous immature erythrocytes
    3. Hemorrhage in eyes and base of fins
E. Histopathology:
    1. Multifocal areas of necrosis of the myocardium, liver, kidney, and posterior gut (leading to cast formation)
    2. Syncytial cells involving the acinar cells of the pancreas is considered to be a pathognomonic sign.
F. Transmission of the virus is believed to be direct.

G. Control is by avoiding exposing susceptible trout to the virus. If the disease occurs, raising the water temperature to 15°C or more will minimize losses.

## CHANNEL CATFISH VIRUS

A. Herpesvirus
B. Observed in fry or fingerling channel catfish (less than 10-gram weight) during the summer when water temperatures are above 22°C.
C. Clinically these fish usually show erratic swimming or spiraling followed by terminal lethargy. Mortality is very high.
D. Lesions:
   1. Hemorrhage at the base of the fins and skins;
   2. Ascites; exophthalmos; and pale gills;
   3. Kidneys swollen and pale with hemorrhage;
   4. Spleen is enlarged and dark red;
   5. Gills usually pale;
E. Histopathology: Multifocal areas of necrosis and hemorrhage are observed in the posterior kidney, liver, intestines, and spleen.
F. Infection is direct with transmission of the virus in the water or feed. Piscivorous birds, snakes, or turtles may mechanically carry the virus from pond to pond. Transovarian transmission has not been conclusively demonstrated but is suspected. Survivors are persistently infected and become carriers for life.
G. Control of the disease is by sanitation, purchasing of virus free brood stock and lowering water temperature to less than 19°C during an outbreak to lessen the mortality.

## EPITHELIOMA PAPILLOSUM (FISH POX)

A. Herpesvirus cyprini (Cyprinid herpesvirus 1)
B. Non-fatal disease is observed in carp and other cyprinids
C. Lesions: Elevation of the epidermis with the formation of white to yellow plaques over the body of the fish. Healed lesions usually turn black.
D. Histopathology: There is epidermal hyperplasia with the epithelial cells occasionally demonstrating intranuclear inclusion bodies.
E. Transmission is unknown, however, it is probably direct.

## INFECTIOUS HEMATOPOIETIC NECROSIS (IHN)

A. Rhabdovirus
B. The disease is observed in the fry of trout (rainbow) and salmon (Chinook and sockeye) with mortality up to 100%.
C. Clinical signs and lesions:
   1. Fish become lethargic or hyperactive.
   2. The fish become dark due to increase in pigmentation.

3. Exophthalmus, abdominal distension, and fecal cast.
4. Hemorrhage on skin and viscera primarily at base of fins, behind the skull, and above the lateral line.
5. Anemia with pale gills.
6. Surviving fish may develop scoliosis.

D. Histopathology: There is prominent necrosis of hematopoietic tissue including melanomacrophages of the kidney, red pulp of the spleen and hepatic parenchyma. Necrosis of the submucosal eosinophilic granular cells is considered pathognomonic for IHN. (This lesion is observed in other systemic viral diseases.) Intranuclear and intracytoplasmic inclusions are occasionally observed in acinar and islet cells of pancreas.

E. The virus is transmitted by direct contact with infected survivors or by feeding contaminated feed. The virus is probably shed in contaminated semen and eggs. The disease is most severe at 10oC and rare at temperatures above 15°C.

## VIRAL HEMORRHAGIC SEPTICEMIA

A. Rhabdovirus

B. Widespread and very contagious viral disease of rainbow trout. This is a serious disease of trout in Europe. Affects both salmonids in fresh water and seawater. Disease occurs in temperatures below 14°C.

C. Two forms of the Disease --- Acute and Chronic

1. Acute disease: High mortality in affected fish. Fish have pale gills, dark body coloration, ascites, exophthalmus and erratic swimming behavior (spiraling). Hemorrhage is a common finding in the eyes, skin, serosal surfaces of the intestines and muscles. Necrosis of the hematopoietic and lymphoid elements of the anterior kidney and congestion and necrosis of the hepatic parenchyma are histopa-thologic findings.
2. Chronic disease: See a slower prolonged mortality. Fish become lethargic, have pale anemic gills, darken skin coloration, exophthalmus, and distention of the abdominal cavity. Internal organs are commonly involved with splenomegaly, hepatomegaly, and swollen kidneys.
3. Turbot, sea bass, and Atlantic salmon are commonly affected by similar viruses.

D. Transmission is believed to be direct with contact of carriers and contaminated water and feed. Vertical transmission via the egg is not reported.

## SPRING VIREMIA OF CARP (SVC) AND SWIM BLADDER INFECTION VIRUS (SBI)

A. Caused by several subtypes of Rhabdovirus carpio.

B. Disease occurs in carp and other cyprinids.

C. Clinical Signs and Lesions:

1. Loss of coordination and equilibrium.
2. Exophthalmus and abdominal distension (ascites).
3. Inflamed and swollen vent.
4. Edema and hemorrhage in many organs.
5. In SBI see pronounced inflammation and hemorrhage of swim bladder.

D. Transmission: Virus shed in feces and found in contaminated eggs.

## INFECTIOUS PANCREATIC NECROSIS (IPN)

A. Birnavirus

B. Affects most salmonids primarily rainbow trout and brook trout. IPN has also been implicated in disease among several nonsalmonid fish.

C. Clinical signs and lesions:

1. IPN is characterized by a sudden explosive outbreak with high mortality.
2. Affected fish become dark and rotate their bodies while swimming.
3. Diseased fish usually have distended abdomens and exophthalmus.
4. The presence of a gelatinous material in the stomach and anterior intestine is highly suggestive of IPN; mucoid fecal casts are common.
5. Infected fish commonly have a low hematocrit and hemorrhage in gut, primarily in the area of the pyloric ceca.

D. Histopathology: Histologically, there is necrosis of the pancreatic acini, gut mucosa, and renal hematopoietic elements. A moderate inflammatory infiltrate is usually observed around the pancreatic acini. Hyalin degeneration of skeletal muscle is also observed.

E. Virus can be transmitted vertically in the eggs.

## FUNGUS IN FISH (SAPROLEGNIA)

Fungus is as the name tells us caused by fungi (Saprolegnia) and will only attack weakened fish that has already been infected by another disease. Fungus will never attack an otherwise strong and healthy fish and the best way of avoiding this disease it therefore to keep your aquarium clean and your fish healthy and well fed.

Fungus will eventually kill infected fish when it has spread through the fish even if this often takes quite a while to happen. The symptoms of fungus are gray or white growth on the fins and skin of the fish. The gray or white areas grow larger and more cotton like as the disease progresses.

Fungus is fortunately rather easy to treat and there are several different ways to treat it. One is to add methylene blue to the water. Methylene blue are most commonly used to prevent fungi on fish eggs but can be used on adult fish as well even if it isn’t necessarily the best option for adult fish.

Personally I recommend using methylene blue for treating and prevent fungi on eggs and fry. On adult fish I instead recommend using a phenoxethol solution or commercial antifungal medicine designed to treat fungus in fish. If you choose to use phenoxethol solution you should add 10ml of 1% phenoxethol solution per litre aquarium water.

The treatment can be repeated after a few days if needed but should not be done more than twice since repeated use can have damaging side effects. Severely infected fish can be caught and swabbed with a povidone, iodine or mercurochrome soaked cloth to increase the effectiveness of the treatment. Make sure to increase the water quality and fish health in your aquarium once the fungus has been cured to avoid the problem in the future.

## VIRAL HEMORRHAGIC SEPTICEMIA

Viral hemorrhagic septicemia (VHS) is a deadly infectious fish disease caused by the Viral hemorrhagic septicemia virus (VHSV, or VHSv). It afflicts over 50 species of freshwater and marine fish in several parts of the northern hemisphere. VHS is caused by the viral hemorrhagic septicemia virus (VHSV), different strains of which occur in different regions, and affect different species. There are no signs that the disease affects human health. VHS is also known as "Egtved disease," and VHSV as "Egtved virus."

Historically, VHS was associated mostly with freshwater salmonids in western Europe, documented as a pathogenic disease among cultured salmonids since the 1950s. Today it is still a major concern for many fish farms in Europe and is therefore being watched closely by the European Community Reference Laboratory for Fish Diseases.

It was first discovered in the US in 1988 among salmon returning from the Pacific in Washington State. This North American genotype was identified as a distinct, more marine-stable strain than the European genotype. VHS has since been found afflicting marine fish in the northeastern Pacific Ocean, the North Sea, and the Baltic Sea. Since 2005, massive die-offs have occurred amongst a wide variety of freshwater species in the Great Lakes region of North America.

### Virus taxonomy

VHSV is a negative-sense single-stranded RNA virus of the order Mononegavirales, family Rhabdoviridae, and genus Novirhabdovirus. (ICTV virus classification, unlike biological classification, begins with the order.) Its NCBI-assigned Taxonomy ID is 11287. Another related fish rhabdovirus in the Novirhabdovirus genus is the Infectious hematopoietic necrosis virus (IHNV), which causes infectious hematopoietic necrosis (IHN) disease in solmonidae.

The viral cause of the disease was discovered in 1963 by M. H. Jenson. The virus is an enveloped, bullet-shaped particle, about 180 nm long by 60 nm in diameter, covered with 5 to 15 nm long peplomers.

### Molecular Virology

The genome of VHSV is composed of approximately 11-kb of single stranded RNA, which contains six genes that are located along the genome in the 3?-5?

order: 3?-N-P-M-G-NV-L-5?, nucleocapsid protein (N), polymerase-associated phosphoprotein (P), matrix protein (M), surface glycoprotein (G), a unique non-virion protein (NV), and virus polymerase (L).

Reverse genetics is a powerful tool to study and characterize the previously unknown viral genes. Reverse genetics system is currently available for VHSV. A vaccinia virus free reverse genetic system for Great Lakes VHSV (Genotype IVb) was developed by a research group from the USA. This system allows the investigators to explore the functional properties of individual viral genes of VHSV in detail.

This system was immediately utilized to characterize the non-virion (NV) gene of Novirhabdoviruses. Even though it has been demonstrated that the NV gene is not necessary for viral replication, it is highly essential for viral pathogenesis. A new role of NV protein has been discovered and demonstrated that it inhibits apoptosis at the early stage of viral infection. This discovery unlocked the mystery of presence of NV protein in Novirhabdoviruses.

## Virus subtypes

Different isolates (unique strains) of VHSV are typically grouped by genotyping. It is found that genotype groups are divided more geographically than by host species. Earlier studies used different numbering systems, but the following system has come into common usage based on genotype similarity based on sequencing of the N- and G-genes. Types I-III are enzootic to Europe, and Type IV to North America, and Type I and type IV isolates are further subdivided, as follows:

Type I-a was the only strain known from VHSV's discovery in 1963 until the late 1988, isolated to fish farms in continental Europe, affecting primarily rainbow trout and occasionally brown trout or pike.

In 1988, the first marine strain of VHSV, now designated type IV, was found in normal-appearing salmon returning from the Pacific to rivers of Washington State. This strain and other marine strains was not lethal to rainbow trout.

The discovery prompted further studies, and by the mid-1990s, marine VHSV was found in eight species along the northern North America's Pacific coast, and 14 species in and around the Atlantic's North Sea. 1996 saw the first VHSV in Japan, among Japanese flounder farmed in the Seto Inland Sea, and different genotypes have appeared in different areas since then.

Type IV was later found off North America's northern Atlantic coast, in Atlantic herring (Clupea harengus) mummichog (Fundulus heteroclitus), stickleback (Gasterosteus aculeatus aculeatus), brown trout (Salmo trutta), and striped bass (Morone saxatilis), as well as dozens of freshwater species in the Great Lakes.

VHSV continues to be found in new geographical areas, in new species of fish. This is thought to represent both the spread of the virus into new areas, as with VHSV egg and live fish transfers from North America to Asia, or feeding of raw marine fish to inland farmed trout in Finland, as well as discovery of existing populations, as with an apparently well established marine reservoir in the Black Sea.

To keep track of the distribution of different VHSV genotypes, a database called Fishpathogens.eu has been created to store data on different fish pathogens (including VHSV) and their sequences.

## GREAT LAKES TYPE IV-B

The Type IV-b strain of VHSV has been spreading among freshwater fish in the Great Lakes region since at least 2003, resulting in some massive die-offs since 2005 of many species in the affected lakes. Originally found off the Atlantic coast of Canada, it was considered a low mortality marine strain. Its first detection in freshwater was in Lake Ontario in 2005, and then in an archived 2003 sample from Lake St. Clair. The isolate was named MI03GL and was sequenced for its entire genome.

The North American genotype of the virus, in addition to moderate mortality to salmonid species, including various varieties of trout, is also proving virulent among a wide variety of warm-water species previously considered resistant to VHS. The Great Lakes region variant has killed lake trout, steelhead trout, chinook salmon, yellow perch, gobies, emerald shiners, muskies, whitefish, and walleye. While the European strain of VHSV is particularly deadly to rainbow trout, the Great Lakes region variant affects the species only mildly, as is typical with primarily marine genotypes.

### Great Lakes regional distribution

An archived 2003 sample from Lake St. Clair of Great Lakes muskellunge is the earliest confirmed case of VHSV within the Great Lakes region. Lake St. Clair connects to Lake Erie through the Detroit River to the south, and to Lake Huron through the St. Clair River to the north. The sample was not tested for VHS until 2005, after the disease was detected in Lake Ontario.

2005 samples of Lake Ontario freshwater drum and Lake Huron lake whitefish were infected with VHS. Initially classified as an unknown rhabdovirus, the Lake Ontario sample was confirmed to be VHSV in 2006, while the Lake Huron sample was confirmed to be VHSV in 2007.

2006 saw mass die-offs from VHS in Lake Erie, the St. Clair River, the Detroit River, and the St. Lawrence River, which connects the Great Lakes to the Atlantic Ocean. VHS was further detected in the Niagara River, which connects Lake Erie to Lake Ontario. It was also found in a walleye die-off in the landlocked inland Conesus Lake, the westernmost of the Finger Lakes in western New York state. This was the first case in the region outside waters contiguously connected to the Great Lakes.

On May 12, 2007, the Wisconsin DNR announced the likely presence of VHS in Wisconsin's inland Little Lake Butte des Morts. Preliminary tests of samples of freshwater drum collected on May 2 were positive, and the announcement came amidst a die-off of hundreds of freshwater drum there and on neighboring Lake Winnebago. Preliminary tests later indicated VHS in specimens from Lake Winnebago. The lakes drain through the Fox River to Lake Michigan's Green Bay.

On May 17, 2007, the Michigan DNR confirmed the presence of VHS in the Michigan's inland Budd Lake, a popular fishing destination is in the central part of Michigan's lower peninsula. A major die-off of VHS-positive muskies, bluegills, and black crappie began on April 30, 2007.

On May 24, 2007, preliminary tests indicated the presence of VHS in a brown trout from Lake Michigan, the second largest freshwater lake in the United States. Contamination in the lake had been expected for months by experts, since the presence of VHS was confirmed in the connected Lake Huron.

On July 14, 2007, federal labs confirmed the presence of VHS in Skaneateles Lake, the second of New York's Finger Lakes to test positive for the disease. The disease caused a large die-off of bass in the spring of 2007.

## Government regulation

Before the Type IVb die-offs in the Great Lakes, few states regulated transfer of live species other than for salmonids. Since 2005, new policies have been adopted concerning fish and egg transfer, use of live bait, and water transfer, aimed at curtailing the spread to new lakes and rivers in the region.

As of July 13, 2007, new rules have been enacted in the Canadian province of Ontario, and US states of Michigan, New York, Ohio, Pennsylvania, and Wisconsin, while they are currently being drafted in Illinois, Indiana, and Minnesota. Additionally, the USDA's Animal and Plant Health Inspection Service issued a federal order in the fall of 2006 barring the transfer of all live susceptible species from the eight states bordering the lower Great Lakes, as well as importing such species from The Canadian provinces of Ontario and Quebec.

## Transmission

VHSV can be spread from fish to fish through water transfer, as well as through contaminated eggs, and bait fish from infected waters. The emerald shiner is a particularly popular bait fish in the Great Lakes region, and is among the species afflicted.

Survivors of the disease can become lifelong carriers of the virus, contaminating water with urine, sperm, and ovarian fluids. The virus has been shown to survive two freeze/thaw cycles in a conventional freezer, suggesting both live and frozen bait could be a transmission vector. In Europe, the gray heron has spread the virus, but it does so mechanically; the virus is apparently inactive in the digestive tract of birds.

## Symptoms

Fish that become infected experience hemorrhaging of their internal organs, skin, and muscle. Some fish show no external symptoms, but others show signs of infection that include bulging eyes, bloated abdomens, bruised-looking reddish tints to the eyes, skin, gills and fins. Some infected fish have open sores that may look like the lesions from other diseases or from lamprey attacks.

There may also be a nervous form of the disease where fish are constantly flashing and showing abnormal behaviour.

## DIAGNOSIS

### Field diagnosis

Living fish afflicted with VHS may appear listless or limp, hang just beneath the surface, or swim very abnormally, such as constant flashing circling due to the tropism of the virus for the brain.

External signs may include darker coloration, exophthalmia ("pop eye"), pale or red-dotted gills, sunken eyes, and bleeding around orbits (eye sockets) and at base of fins.

Genetics researchers at the Lake Erie Research Center at the University of Toledo are developing a test that will speed diagnosis from a month to a matter of hours.

### Gross pathology (Non-laboratory)

VHSV is a hemorrhagic disease, meaning it causes bleeding. Internally, the virus can cause petechial hemorraging (tiny spots of blood) in internal muscle tissue, and petechial or severe hemorrhaging in internal organs and other tissues. Internal hemorrhaging can be observed as red spots inside a dead fish, particularly around the kidney, spleen, and intestines, as well as the swim bladder, which would normally have a clear membrane. The liver may be pale, mottled with red hyperemic areas, the kidney may be swollen and unusually red, the spleen may be swollen, and the digestive tract may be empty.

External signs are not always present, but if they are, hemorrhaging on the skin's surface can appear as anywhere from tiny red dots (petechiae) to large red patches.

### Histopathology (Microscopic tissue analysis)

Preliminary diagnosis involves histopathological examination, observing tissues through a microscope. Most tissue changes can be observed as minor to major necrosis (cell death) in the liver, kidneys, spleen, and skeletal muscle. The hematopoietic (blood-forming) areas of the kidney and spleen are the initial area of infection, and should show necrosis. The gill may have thickened lamellae, and the liver may have pyknotic nuclei. Skeletal muscle accumulates blood but does not suffer much damage.

## Virology (Definitive testing)

Electron microscopy can reveal the bullet-shaped rhabdovirus, but is not adequate for definitive diagnosis.

The Manual or Diagnostic for Aquatic Animals, 2006, is the standard reference for definitive tests. In most cases, cell culturization is recommended for surveillance, with antibody tests and reverse transcription polymerase chain reaction (RT-PCR) and genetic sequencing and comparison for definitive confirmation and genotype classification.

Virus neutralisation is another important method of diagnosis, especially for carrier fish.

## Prevention

Thoroughly cleaning boats, trailers, nets and other equipment when traveling between different lakes and streams also helps. The only EPA-approved disinfectant proven effective against VHS is Virkon AQUATIC (made by Dupont). Chlorine bleach kills the VHS virus, but in concentrations that are much too caustic for ordinary use. Disinfecting stations can be found at various inland lake boat launches in the Great Lakes region.

# PATHOGENESIS OF ACUTE VIRAL DISEASE INDUCED IN FISH

A lethal disease of koi and common carp (species Cyprinus carpio) has afflicted many fish farms worldwide since 1998, causing severe financial losses. Morbidity and mortality are restricted to common carp and koi and appear in spring and autumn, when water temperatures are 18 to 28°C. We have isolated the virus causing the disease from sick fish, propagated it in koi fin cell culture, and shown that virus from a single clone causes lethal disease in carp and koi upon infection.

Intraperitoneal virus injection or bathing the fish in virus-containing water kills 85 to 100% of the fish within 7 to 21 days. This virus is similar to the previously reported koi herpesvirus; however, it has characteristics inconsistent with the herpesvirus family, and thus we have called it carp interstitial nephritis and gill necrosis virus. We examined the pathobiology of this disease in carp by using immunohistochemistry and PCR. We found large amounts of the virus in the kidneys of sick fish and smaller amounts in liver and brain.

A rapid increase in the viral load in the kidneys was detected by using both immunofluorescence and semiquantitative PCR. Histological analyses of fish at various times after infection revealed signs of interstitial nephritis as early as 2 days postinfection, which increased in severity up to 10 days postinfection. There was severe gill disease evidenced by loss of villi with accompanying inflammation in the gill rakers. Minimal focal inflammation was noted in livers and brains. This report describes the etiology and pathology of a recently described viral agent in fish.

Since 1998 many carp and koi farms have been afflicted by a disease with a high mortality rate, resulting in a drastic reduction in production. This disease is caused by a viral agent and has been observed in many farms in the northern and western regions of the United States, Europe, Israel, Indonesia, Japan, and Korea.

The rapid spread of this disease is probably due to the intensive worldwide trade of these splendid fish, mostly without veterinary supervision. The disease appears in open-air ponds during spring and fall, with clinical signs including fatigue, gasping movements in shallow water, gill necrosis, sunken eyes, pale patches on skin, and increased mucus secretion. Studies performed under laboratory conditions confirmed that the virus induces the disease only at the permissive temperature of 18 to 28°C.

Although the disease is highly contagious, its morbidity and mortality are restricted to koi and common carp populations. Other fish of the tilapia group (Oreochromis niloticus), silver perch (Bidyanus bidyanus), silver carp (Hypophthalmichthys molitrix), gold fish (Carassius auratus), and grass carp (Ctenopharyngodon idella) were found to be fully resistant to the disease, even after long cohabitation with sick fish at the permissive temperature.

This disease has been described previously by Hedrick and coworkers, who isolated the causative pathogen, the koi herpesvirus (KHV). In agreement with previous studies of KHV, the virus that we isolated and designated carp interstitial nephritis and gill necrosis virus (CNGV) has an icosahedron-shaped core of 100 to 110 nm, is an enveloped virus, and bears thread-like structures (tegument) on the core surface resembling those of the herpesvirus.

Although we believe KHV and CNGV to be the same, we have found that CNGV contains an asymmetrical electron-dense region within the viral core, which is probably the genomic nucleoprotein complex, and that its genome is a double-stranded DNA molecule of 270 to 290 kbp (Hutoran et al., submitted), which is larger than those of all known Herpesviridae members.

Since the viral genome sequences determined so far, including part of the thymidine kinase gene (GenBank accession numbers AY208988 to -91, AJ535112, and AF411803), contain only small fragments (16 to 45 bp), similar to the case for other DNA viral genomes, we believe that it is too early to classify the virus phylogenetically. Thus, although CNGV may be similar or identical to KHV, we believe that it is preferable to designate it according to its pathological manifestation as CNGV.

## Morphology of KHV/CNGV

The morphology of KHV/CNGV and its genomic DNA sequences have been intensively studied. However, there is little information about its effect in tissue culture, the pathogenesis of the disease, target organs of the virus, and the way it spreads in fish organs. This study was undertaken to elucidate these important points. Here we show that (i) CNGV propagates and induces severe cytopathic effects at 3 to 5 days postinfection (p.i.) in fresh koi fin cell (KFC) cultures; (ii) cloned CNGV harvested from KFC cultures induces the same disease upon inoculation of healthy koi and common carp, with a mortality rate of 75 to 95%; (iii) the amounts of CNGV DNA and specific viral antigens in the kidney and blood increase during the first

7 days p.i.; and (iv) the virus induces progressive pathogenic effects in the kidney, liver, and gills, as shown by histological and immunohistochemical examinations.

## MATERIALS AND METHODS

### Fish

Common carp with an average weight of 50 g (approximately 4 months old) were grown in 100- or 500-liter tanks. The water temperature was maintained at 22 to 24°C, and fresh water was supplied at 0.5 or 0.9 liter/min.

### Fish infection

Infection was carried out by cohabitation, bathing, or injection.

(i) Cohabitation : Two symptomatic fish from the stock of sick fish continuously maintained at Dor Research Station were added to a 500-liter tank containing 30 healthy fish and left for 24 h.

(ii) Bathing : Healthy fish were kept in a 20-liter tank, to which virus was added to achieve a final concentration of ?30 PFU/ml. After 50 min under these conditions, the fish were transferred into large tanks. The control group was immersed in water containing medium harvested from uninfected KFC cultures under the same conditions.

(iii) Injection : A 0.2-ml volume of culture medium containing 100 PFU of isolated virus was injected intraperitoneally into healthy fish.

### Cell cultures

Cultures were prepared as previously described by Hasegawa et al. and Neukirch et al. Briefly, caudal fins were removed from 50 g of koi fish under anesthesia, bathed in 1% sodium hypochloride solution for 1 min, and rinsed in 70% ethyl alcohol for a few seconds. The fins were then washed three times for 0.5 min each in phosphate-buffered saline (PBS) containing penicillin and streptomycin.

They were transferred to petri dishes and extensively minced with scissors, and small semidry tissue pieces of approximately 1 mm3 were placed in dry 50-ml culture flasks (Nunc). After 60 min of incubation at room temperature, the clumps adhering to the flasks were covered with culture medium containing 60% Dulbecco's modified Eagle's medium, 20% Leibovitz (L-15) medium, 10% fetal calf serum (Biological Industries, Kibbutz Beit Haemek, Israel), and 10% tryptose phosphate (Difco) and supplemented with 1% HEPES and antibiotics.

Cells grew to form a monolayer over a period of 10 to 14 days in a 22°C incubator supplemented with 5% CO2. The monolayer cultures were trypsinized and transferred into new flasks with fresh medium.

### Purification of virus from culture medium

Medium harvested from infected KFC cultures was cleared of cells and cell debris by centrifugation for 10 min at 10,000 × g. The virus was then pelleted by centrifugation in a Beckman Ti-60 rotor for 50 min at 100,000 × g. Pellets were suspended in PBS and loaded on a 15 to 65% (wt/vol) sucrose gradient prepared in PBS and centrifuged for 60 min at 110,000 × g in a Beckman SW28 rotor. Bands were aspirated from tubes, diluted 10-fold in PBS, and repelleted. The pellets were suspended in PBS and frozen at ?70°C for further investigation.

## Antibodies

Anti-CNGV serum was generated by immunizing a rabbit with 0.1 mg of purified CNGV emulsified 1:1 in Freund's complete adjuvant. The rabbit was boosted three additional times at 10- to 14-day intervals with 0.05 mg of purified CNGV mixed 1:1 with Freund's incomplete adjuvant and bled three times between 7 and 10 weeks after the first immunization to collect antiserum. In order to reduce nonspecific background, the antisera were absorbed into a mixture of KFCs and koi fish powder prepared from muscles and kidneys of healthy fish, as described by Harlow and Lane.

## Indirect immunofluorescence microscopy

At 16 days after cohabitation, sick fish were anesthetized and their kidneys, livers, and brains were removed and used for the preparation of touch imprint slides. Organs taken from healthy fish were used as controls. These were fixed for 10 min in 100% methanol, washed with PBS, and blocked by incubation for 60 min with low-fat milk containing 50% fetal calf serum.

The slides were then incubated for 60 min with rabbit preimmune serum or anti-CNGV serum (diluted 1:10,000 in PBS), washed with PBS, incubated for 60 min with fluorescein isothiocyanate-conjugated swine anti-rabbit antibodies (Dako), washed with PBS, and analyzed with a fluorescence microscope equipped with a 40× objective.

## Histological analysis

For histological analysis, brain, gill, liver, and kidney tissue samples were fixed overnight in 4% buffered formaldehyde and embedded in paraffin. Five-micrometer sections were stained with hematoxylin and eosin. Selected sections were also stained with Gram, periodic acid-Schiff, and Ziehl-Neelsen stains.

## Immunohistochemistry

Fish infected by intraperitoneal injection of 200 PFU of CNGV per ml were used for immunohistochemistry. Formalin-fixed, paraffin-embedded tissue samples were cut into 5-?m sections, deparaffinized in xylene, and rehydrated through a series of graded ethanol solutions. Sections were incubated for 1 h with polyclonal rabbit anti-CNGV serum diluted at 1:30,000 in CAS block (Zymed Laboratories) at room temperature.

Antibodies were detected by using the Envision+ kit (DAKOCytomation, Glostrup, Denmark) according to the manufacturer's instructions. 3-Amino-9-ethylcarbazole was used as the chromogen. As a control, sections were incubated with nonimmune rabbit serum at the same dilution. For competition experiments the antiserum was preincubated with purified virus or BSA for 30 min before application to the tissue sections.

### Electron microscopy

For ultrastructural studies, kidneys from healthy or infected fish were removed, dissected, rapidly immersed in 2.5% glutaraldehyde in cacodylate buffer, postfixed in osmium tetraoxide, and embedded in Epon. Ultrathin sections were cut with an Ultracut E microtome (Reichert-Jung, Vienna, Austria). Sections were stained with lead citrate and uranyl acetate and viewed by transmission electron microscopy (Philips 420 electron microscope at 80 kV).

## RESULTS

### Cloned CNGV is sufficient to induce mortal disease in carp

Virus was isolated from sick fish by cocultivation of KFCs with cells from kidney of sick fish, and aliquots were used to infect KFC cultures. The virus isolated from sick fish produces typical plaques on KFCs at 4 to 6 days p.i. Cytopathic effect appears in KFC cultures at 3 to 5 days p.i. At 2 to 3 days p.i. the volume of the infected cells increases and abundant endoplasmic vacuoles appear in the cells.

At a later stage, cells become round, die, and detach from the substrate. Significant amounts of virus are released into the culture medium at 3 to 5 days p.i. In order to determine whether a virus isolated from a single plaque (cloned virus) is sufficient to induce the carp disease, we infected KFC cultures with decreasing titers of the virus. Following absorption, the cultures were overlaid with agar and incubated at 22 to 23°C for 5 days.

The number of plaques decreased proportionally to the virus dilutions, displaying a one-hit curve, demonstrating that a single virus is sufficient to generate each plaque (results not shown). Virus from a single plaque was propagated, purified, and injected into healthy fish as described in Materials and Methods, and 91% of fish injected with ?102 PFU of the cloned virus died within 21 days following infection.

All the diseased fish showed the classic clinical symptoms of CNGV infection, including gasping movements, sunken eyes, and swimming in shallow water. In addition, we randomly sampled four fish at 2, 4, 6, 8, and 10 days p.i. for histological analysis. All infected fish showed typical changes of CNGV infection (see below). Control fish injected with medium harvested from uninfected cultures remained healthy throughout the 30-day experiment.

The cloned virus was also used to infect fish by bathing as described in Materials and Methods. Ninety-one percent of the fish exposed to virus in the water died between 10 and 24 days p.i., while all of the control fish remained healthy during the entire experimental period. These results demonstrate that the virus can be transmitted by injection or via water and that virus cloned from a single plaque is sufficient to induce the carp disease.

## Identification of CNGV DNA in blood and kidneys of infected fish.

To follow the kinetics of the virus's appearance, healthy fish (50 ± 10 g) were cohabitated with sick fish or bathed for 1 h in water containing CNGV at a final concentration of 30 PFU/ml. Noninfected fish kept under identical conditions served as a control group. Five fish from each group were randomly sampled at each time point.

Using semiquantitative PCR, we found that the amount of viral DNA in extracts of both kidney and blood cells from fish infected by cohabitation or bathing increased throughout the testing period. CNGV DNA was detected at low levels in blood and kidney as early as 1 day after exposure of fish to the virus by bathing, whereas it was not detected until day 3 or 5 in kidney or blood cells of fish infected by cohabitation and was absent in kidney and blood of uninfected fish.

## Identification of CNGV in fish organs.

To search for viral protein in fish organs, we generated a rabbit anti-CNGV antiserum. The rabbit anti-CNGV serum bound to touch imprints of kidney taken from symptomatic fish at 16 days p.i. by cohabitation but not to those from uninfected fish, indicating that the virus is present in kidneys of infected fish.

We also used immunofluorescence with rabbit anti-CNGV serum to assess the presence of virus in the brains and livers of sick fish. Staining in touch imprints from these organs was significant but was lower than that found in kidney imprints. Preimmune rabbit serum did not bind to touch imprints from any of these organs, demonstrating the specificity of the interaction (data not shown).

These experiments suggest that viral proteins are extensively expressed in kidney of diseased fish and less so in the brain and liver. These results are in agreement with our experiments isolating virus from diseased fish organs (results not shown) and with our semiquantitative PCR experiments, both of which show that the kidney harbors large quantities of infectious virus.

## Histopathology

To further characterize the disease process, we infected fish by bathing and analyzed the histopathological changes induced by viral infection in several organs at various time points following infection. Pathological changes were noted in the gills as early as 2 days p.i., as evidenced by a loss of lamellae accompanied by a mixed inflammatory cell infiltrate. However, these changes were not present in all of the filaments, and considering the frequent involvement of the gills in numerous inflammatory processes (T. T. Poppe, Histopathology workshops: notes and images, EAFP Histology Workshop, 2001), these findings are not diagnostic at this early stage of infection.

From 6 days p.i. onward, changes in the gills became more pronounced, with complete effacement of the gill architecture accompanied by severe inflammation in nearly all of the filaments. Congestion of the gills' central venous sinus was also evident at this stage. Similar changes were evident at 8 and 10 days p.i. Many infectious diseases induce a pronounced effect in the gill filaments.

Interestingly, in fish infected with CNGV, the gill rakers showed effects that were more easily recognizable than the changes observed in the filaments. These include increased subepithelial inflammation and congestion of the blood vessels in the gill arch. The inflammatory process was accompanied by attenuation of the rakers' height. As early as 6 days p.i., focal sloughing of the surface epithelium was noted. This was even more marked at later time points assayed (not shown).

While gill injury precedes all other histological changes, it is plausible, as argued before, that this effect is caused by secondary infections. To assess whether bacterial, fungal, or parasitic infectious agents are involved in this process, we stained histological sections of gills at different time points with Gram, periodic acid-Schiff, and Ziehl-Neelsen stains, which facilitate the detection of microorganisms.

No increase in the amount of microorganisms was detected on days 2 to 6 following infection. This is in contrast with the marked inflammatory response in the gills that is evident as early as 2 days p.i. Only at 8 and 10 days p.i., when focal denudation of the gill epithelium occurs, were increased numbers of microorganisms noted (data not shown), suggesting that the primary gill disease induced by CNGV predisposes the fish to secondary infections.

In addition to the gills, the most prominent pathological changes were noted in the kidneys. A mild peritubular inflammatory infiltrate was evident as early as 2 days p.i. On day 6, a heavy interstitial inflammatory infiltrate was observed, along with congestion of blood vessels. At 8 days p.i. the infiltrates were more severe and were accompanied by a feathery degeneration of the tubular epithelium in many nephrons, together with the presence of intraepithelial lymphocytes.

As early as 6 days p.i., large cells with a foamy distended cytoplasm and a few intranuclear inclusion bodies were scattered among the inflammatory interstitial cells. These "foamy" cells are reminiscent of the cytopathic effect observed in infected cultured cells. Liver analysis showed mild inflammatory infiltrates, located mainly in the parenchyma, while brain sections showed focal meningeal and parameningeal inflammation.

## Detection of CNGV in fish tissues by immunohistochemistry.

As noted above, both PCR and immunofluorescence detect viral particles in kidneys, with the latter method also showing that such particles are present in brains and livers of diseased fish. To verify these findings and to identify the specific cells in which viral proteins are expressed, we performed immunohistochemical analyses of kidneys, brains, livers, spleens, and gills from infected and healthy fish. Kidneys from healthy fish showed no staining with anti-CNGV antiserum.

As early as 2 days p.i., viral proteins were detected in the kidney by the rabbit anti-CNGV serum, in conjunction with the appearance of the lymphoid inflammatory infiltrate in the renal interstitium (data not shown). At this stage of infection, stained cells were detected only in the interstitium, most of them large and corresponding to the foamy cytomegalic cells described above.

The number of immunostained interstitial cells was maximal at 6 days p.i. and remained constant through 10 days p.i. Interestingly, at 10 days p.i., when hematoxylin and eosin staining shows maximal feathery degeneration of the tubular epithelium, viral proteins were also detected in some of the tubular epithelial cells This finding suggests that the virus is not restricted to cells of the hematolymphoid lineage.

No virus-positive cells were detected in a serial section from each kidney that was immunostained with preimmune rabbit serum and served as a negative control. To further verify the specificity of the immunohistochemical method, we preincubated the antiserum with purified virus. Preincubation with BSA served as a control. show serial sections of a kidney from an infected fish at 8 days p.i. that were incubated with untreated anti-CNGV antiserum or with anti-CNGV antiserum preincubated with 0.1, 1, or 10 ?g of viral proteins.

Preincubation with increasing amounts of viral protein decreased the staining intensity and, at high concentrations, abolished staining. In contrast, preincubation of the antiserum with BSA at the same concentrations did not decrease the staining intensity.

Immunohistochemistry with anti-CNGV antibodies showed that the number of infected cells in the gills of diseased fish increased between 2 and 10 days p.i. (data not shown). However, a few positive cells were detected in all healthy fish assayed. This may represent subclinical disease in these fish, cross-reaction of the antiserum with a similar virus that is commonly present in carp gills, or, alternatively, nonviral proteins present in gills that nonspecifically reacted with our antiserum.

## Detection of CNGV in fish tissues by electron microscopy.

To better characterize the cells which are infected by CNGV, sections from infected kidneys were analyzed by electron microscopy. Viral particles were detected in the cytoplasm of occasional cells in the interstitium of diseased kidneys. The particles had a dark, electron-dense core surrounded by a halo, similar to the case of the particles detected in infected cultured cells and in sections of purified virus preparation (data not shown).

The frequency of virus-containing cells was similar to the frequency of positively stained cells in the immuno-histochemical analysis. The typical virus-bearing cell has an oval, somewhat irregular nucleus contour and abundant cytoplasm. The main cytoplasmic organelles include numerous free ribosomes, a Golgi apparatus, a few mitochondria, and some lysosomes. Neither filaments nor junctional complexes were apparent. Thus, these cells most probably represent monocytes/macrophages. We were unable to detect infected cells in the peripheral blood.

## DISCUSSION

A lethal disease of common carp and koi is caused by a virus here designated CNGV, according to its pathogenic effect in fish. A very similar pathogen was previously isolated from fish suffering from the same symptoms and was designated KHV by Hedrick et al.. The virus that we have isolated is most probably the same virus that was described by Hedrick et al..

However, we believe that CNGV is a more appropriate name, since the virus has several characteristics not found in Herpesviridae. Previous studies have focused on the evolutionary relationship of this virus to other large viruses, by using morphological studies, characterization of the genomic DNA molecule, and sequencing the viral genome.

Although a detailed understanding of the cytopathic effect, the pathogenesis, and the pathogenic effect induced by the virus is extremely important for virus characterization, little is known in this regard. Our report addresses this issue and describes molecular, histological, and immunological findings essential for the characterization of the disease induced by CNGV.

In order to demonstrate that a single plaque-derived clone is sufficient to induce the disease, we first proved that the number of plaques induced by CNGV is directly related to its dilution, i.e., it yields a single-hit curve. Thus, a single infectious unit is sufficient to produce each plaque. We then showed that virus isolated from a single plaque induces the disease in fish and that no helper virus or additional microorganism is required.

The virus isolated from sick fish induced a cytopathic effect and typical plaques in KFCs: it increased the cell volume, induced the formation of abundant cytoplasmic vacuoles in the infected cells, and rendered the cells round shaped before they detached from the substrate. It is noteworthy that infected cells with foamy cytoplasm were also observed in kidneys of diseased fish, suggesting that the production of abundant vacuoles is determined by the virus and not by the target cells.

PCR of DNA from various tissues revealed that viral DNA is present in blood and kidneys of infected fish at very early time points after infection but appears in brain, liver, and spleen only in the late stages of the disease (data not shown). Using semiquantitative PCR, we showed that the viral DNA is present in the kidneys of infected fish. These results are supported by both the immunofluorescence and immunohistochemical studies. Viral DNA was detected in the kidney and blood as early as 3 and 5 days postcohabitation, respectively.

Infection of fish by bathing appears to be more efficient than that by cohabitation, since viral DNA was detected in the kidney and blood as early as 1 day postexposure. However, as in the case of infection by cohabitation, the amount of viral DNA in the kidney began to increase at 3 days p.i. and that in the blood began to increase at 5 or 7 days p.i.

The kidney was found to be the organ in which the virus propagates most efficiently, by both immunohistochemistry and touch imprint techniques.

Interestingly, while immuno-histochemical staining revealed only occasional positive cells, our touch imprint analysis detected many more positive cells. This may be explained in several ways. First, the touch technique samples cells that detach easily from the tissue surface.

It is possible, therefore, that infected white blood cells are more abundant and more prone to detachment than other cell types, yet all cell types are equally represented in the immunohistochemical sections. Second, touch imprints were performed from terminally sick fish at 16 days p.i., while fish used for immunohistochemistry were euthanatized at 6 to 10 days p.i. Finally, formalin fixation, which interferes with antigen-antibody recognition, may result in labeling only of cells bearing very high levels of viral antigens.

At present, we cannot conclusively determine the intracellular localization of the viral propagation. Since this is a DNA virus, one might expect evidence for intranuclear replication. However, electron microscopy, as well immuno-logical detection methods, localized viral particles and/or proteins to the cytoplasm of the infected cells. This interesting aspect of CNGV propagation remains to be solved.

The virus remains infective in water for at least 4 h, explaining the highly contagious nature of the virus in ponds. It is not yet known how the virus enters the fish body, i.e., whether through the gills or through the intestine. Our histological analysis supports the possibility that the virus enters the fish body through the gills, replicates there, and induces mucosal sloughing and necrosis.

It is conceivable that gill injury induced by the virus significantly contributes to fish morbidity. However, host-virus interaction in the gills still needs to be elucidated in order to understand the pathogenic mechanisms responsible for gill injury. The possibility that the virus replicates in the diseased gills and is then shed into the water is in agreement with the rapid and efficient spread of this contagious disease. From the gills, the virus is rapidly transferred to the kidneys, where it resides in white blood cells and induces severe interstitial nephritis. Other organs assayed, including the brain and liver, are relatively spared.

Localization of the virus within white blood cells raises the intriguing possibility that the virus is rapidly transferred to the viscera via infected white blood cells. The virus can then multiply in the kidney and infect the epithelial cells there. Another important question is whether the virus resides in the gills. Despite the technical difficulties mentioned above, the large amounts of virus in this organ suggest that the virus can multiply in the gills.

It can then be released into the water either through shedding or together with the sloughed epithelial and inflammatory cells resulting from severe local inflammation. The ability to invade the fish through the gills, multiply there, and then be released through the water is favorable for the virus. This is analogous to respiratory viruses in mammals that infect the respiratory epithelium, replicate there, and are spread through air droplets and aerosol. This may turn out to be the most common means of spreading of aquatic viruses.

# DISORDERS IN FISHES

## Trimethylaminuria

Trimethylaminuria (TMAU), also known as fish odor syndrome or fish malodor syndrome, is a rare metabolic disorder that causes a defect in the normal production of the enzyme Flavin containing monooxygenase 3 (FMO3). When FMO3 is not working correctly or if not enough enzyme is produced, the body loses the ability to properly break down trimethylamine (TMA) from precursor compounds in food digestion into trimethylamine oxide (TMAO) through a process called N-oxygenation.

Trimethylamine then builds up and is released in the person's sweat, urine, and breath, giving off a strong fishy odor or strong body odor.

## History

The first clinical case of TMAU was described in 1970.

## Symptoms

Trimethylamine builds up in the body of patients with trimethylaminuria. The trimethylamine gets released in the person's sweat, urine, reproductive fluids, and breath, giving off a strong fishy or body odor. Some people with trimethylaminuria have a strong odor all the time, but most have a moderate smell that varies in intensity over time. Individuals with this condition do not have any physical symptoms, and typically appear healthy.

The condition seems to be more common in women than men, for unknown reasons. Scientists suspect that female sex hormones, such as progesterone and/or estrogen, aggravate symptoms. There are several reports that the condition worsens around puberty. In women, symptoms can worsen just before and during menstrual periods, after taking oral contraceptives, and around menopause.

This odor varies depending on many known factors, including diet, hormonal changes, stress level, amount of sweat, other odors in the space, and individual sense of smell.

## Diagnosis

Measurement of urine for the ratio of trimethylamine to trimethylamine oxide is the standard screening test. A blood test is available to provide genetic analysis. The prominent enzyme responsible for TMA N-oxygenation is coded by the FMO3 gene.

A similar test can be used to identify carriers of this condition - those individuals who carry one copy of a mutated gene but do not have symptoms. In this case, the person would be given a high dose of choline (one of the precursors of trimethylamine) and then have their urine tested for elevated levels of trimethylamine.

## Treatment

Currently, there is no known cure or treatment for the disorder. However, some people affected by the disorder live relatively normal lives by managing their symptoms and with counseling. When the condition is suspected or known to occur in a family, genetic testing can be helpful in identifying the specific individuals who have or carry the disorder.

Ways of reducing the fishy odor may include:

- Avoiding foods such as egg yolks, legumes, red meats, fish, beans and other foods that contain choline, carnitine, nitrogen, sulfur and lecithin.
- Taking low doses of antibiotics such as neomycin and metronidazole in order to reduce the amount of bacteria in the gut.
- Using slightly acidic detergent with a pH between 5.5 and 6.5.

Additionally, at least one study has suggested that daily intake of the supplements activated charcoal and/or copper chlorophyllin may improve the quality of life of individuals afflicted with TMAU by helping their bodies to oxidize and convert TMA to the odorless N-oxide (TMAO) metabolite. Study participants experienced subjective reduction in odor as well as objective reduction in TMA and increase in TMAO concentration measured in their urine. The study's success rates varied:

- 85% of test participants experienced complete loss of detectable "fishy" odor
- 10% experienced some reduction in detectable odor
- 5% did not experience any detectable odor reduction

## Genetics

Most cases of trimethylaminuria appear to be inherited in an autosomal recessive pattern, which means two copies of the gene in each cell are altered. The parents of an individual with an autosomal recessive disorder are carriers of one copy of the altered gene. Carriers may have mild symptoms of trimethylaminuria or experience temporary episodes of fish-like body odor.

Mutations in the FMO3 gene, which is found on the long arm of chromosome 1, cause trimethylaminuria. The FMO3 gene makes an enzyme that breaks down nitrogen-containing compounds from the diet, including trimethylamine. These compounds are produced by bacteria in the intestine as they digest proteins from eggs, meat, soy, and other foods.

Normally, the FMO3 enzyme converts fishy-smelling trimethylamine into trimethylamine N-oxide which has no odor. If the enzyme is missing or its activity is reduced because of a mutation in the FMO3 gene, trimethylamine is not broken down and instead builds up in the body. As the compound is released in a person's sweat, urine, and breath, it causes the strong odor characteristic of trimethylaminuria. Researchers believe that stress and diet also play a role in triggering symptoms.

There are more than 40 known mutations associated with TMAU. Loss-of-function mutations, nonsense mutations, and missense mutations are three of the most common. Nonsense and missense mutations cause the most severe phenotypes.

Although FMO3 mutations account for most known cases of trimethylaminuria, some cases are caused by other factors. A fish-like body odor could result from an excess of certain proteins in the diet or from an increase in bacteria in the digestive system. A few cases of the disorder have been identified in adults with liver damage caused by hepatitis.

The evolution of the FMO3 gene has recently been studied, including the evolution of some mutations associated with TMAU.

## NUTRITIONAL DISORDERS IN FISH

Many fishes suffer from nutritional disorders because of a poor diet. Nutritional disorders are the most common reason for sickness and death in aquarium, tank or fishpond fishes.

### Reasons and Prevention

1. Nutritional imbalance in commercial food: Fishes can be either plant eaters (herbivores), meat eaters (carnivores), or both (omnivores). And although commercial food is available for fishes, a nutritional disorder can still occur because each species of fish has a different nutritional requirement, which is not always fulfilled by the commercial food. Therefore, fishes will need more than one type of commercial food to meet their dietary requirement.
2. Incorrectly stored food: Improperly stored food is another reason fishes acquire nutritional disorders. Dry food should be stored in a cool, dry place and replaced after two months.
3. Vitamin deficiency: Nutritional disorders in fishes can also be due to a vitamin deficiency. Vitamin C or ascorbic acid deficiency leads to Broken back disease - where the backbone of the affected fishes get bent (deformed). Vitamin B-complex (thiamin, biotin, niacin, and pyridoxine) deficiency can cause brain, spinal cord and nerve disorders in fishes. Unfortunately, vitamin deficiency is diagnosed only after the fish's death. Therefore, it is important you give your fish a vitamin-rich diet.
4. Infected live food: Food that is alive and infected with bacteria, viruses, fungi and parasites can lead cause problem in your fishes. To prevent such infectious diseases, buy live food only from reputable sources.
5. Feed toxicity: Nutritional disorders caused by toxins found in food occurs frequently in aquarium fishes. The most common of these is the aflatoxin produced by the growth of the mold, Aspergillus flavus, in the stored food. Aflatoxin causes tumors and is fatal in fishes. Store your fish food hygienically and replace it every two months, or when there appears to be mold in it.

## Treatment

Treatment for any fish disorder or disease is difficult. Therefore, it is important you are careful storing food, and that you take any other preventative measures.

## Eating Disorders

You'd think that dieting would be a death sentence for a tiny osteichthyes, no bigger than a bloated paper clip. But some Goby fish see slimming down as survival.

Researchers measured gobies and showed a hierarchy in which each fish had a five percent size difference from the Goby above and below it. When an inferior fish beefed up, its superior would try to drive the power eater out of the group. Thus, the inferior starved as a way out of imminent confrontation. And likely death. (Gobies can't survive alone.)

This fish story doesn't just apply to our marine friends; it bears on human behavior, as well. When I wrote Lying in Weight: the Hidden Epidemic of Eating Disorders in Adult Women, I profiled many couples, in which one partner had an eating disorder.

"Who partners up with a woman weighing 85 pounds?" I asked.

Any one of five categories of men (or women if the male is the one with the eating disorder or the relationship is same-sex). One of these categories fits the Gobie profile well. I called him the, "Macho Man, Control and Conquer."

He is an executive, doctor, attorney, military officer, or minister. He's the man used to managing other people, assigning duties, and being aggressively in charge. He chooses a partner with an eating disorder because she will accept his alpha position, and by extension, her subordinate one. Meanwhile, she starves to shrink into her Stepford role, in part because she feels she cannot exist alone.

This dynamic and others have prompted psychologist Cynthia M. Bulik, PhD, Director of the University of North Carolina Eating Disorders Program, and her colleagues to pilot an intervention for couples in which one member is suffering from anorexia nervosa.

Called, Uniting Couples (in the treatment of) Anorexia Nervosa (UCAN), the intervention starts with the premise that eating disorders are not simply a teen thing. Adults develop anorexia, too. And when they do, they are often in relationships.

Intimacy is tough and tougher still when one partner has an eating disorder. Both partners feel the effects. Therefore, both might benefit from treatment, in this case cognitive-behavioral couple-based interventions that have been successfully employed for the treatment of depression, anxiety disorders, smoking cessation, and cancer.

Back to the fish. While they are not likely to respond to cognitive behavioral therapy, offered by a well-intentioned guppy, fish do construct power imbalance in relationships. Under the sea, these hierarchies help maintain a stable, noncompetitive society.

However, fish are fish and they can maintain smaller sizes without becoming mentally ill.

## BONE AND MUSCLE DISORDERS IN FISH

### Pleistophora Hyphessobryconis & Broken Back

Much like other animals, fishes can have bone and muscle disorders.

### Symptoms and Types

One such bone and muscle disorder is Broken Back Disease, which is typically due to a vitamin C deficiency. This disease will literally bend the fish's backbone. However, injuries are sometimes the cause of an abnormal backbone.

Another typical bone and muscle disorder is caused by the parasite Pleistophora hyphessobryconis. This parasite attacks the skeletal muscle - responsible for movement - of freshwater aquarium fishes, like neon tetra and angelfish. The fishes infected with the parasite, which has caused muscle damage, will then have abnormal muscle movements.

To identify the parasite infection, the veterinarian will perform a microscopic examination. However, there is no treatment for this particular bone and muscle disorder. The only way to prevent the disease from spreading is to remove all infected fishes from the tank, aquarium or fishpond.

### Causes

Bone and muscle disorders in fishes are due to various reasons, including infections, parasites, injuries and nutritional imbalance. The disorders caused by a nutritional imbalance, are because of a deficiency in vitamins (i.e, vitamin C (ascorbic acid), vitamin E and selenium).

### Treatment

It is important you quarantine all fishes that may be infected; this will prevent a possible spread of infection. Also, fishes with bone and muscle disorders due to infection, should be kept in a medicated tank or aquarium - following the treatment prescribed by your veterinarian.

Keep your fishes on a nutritionally balanced diet. Fishes with dietary imbalances should be given vitamins. If caught at an early stage, it will help the fishes suffering from bone and muscle disorders.

## SIGNS OF RENAL DISORDERS

There are few major kidney and urinary tract disorders seen in fishes. Of these the main kidney and urinary tract disorders are Renal Dropsy, Carp-dropsy complex, and Proliferative kidney disease (PKD).

1. Renal dropsy in fishes is caused by the parasite, Sphaerospora auratus. Renal dropsy usually occurs in pond-raised goldfish. There is damage to the kidneys and swelling of the abdomen due to fluid accumulation is the most common sign of renal dropsy. There is no treatment for this kidney disorder and it usually causes death of the infected fish.
2. Carp-dropsy complex is a kidney disorder that usually affects the carp and goldfish. The Carp-dropsy complex disease is caused by the parasite, Sphaerospora angulata. Other complications include viral and bacertial infections, and carp swim-bladder disease. This is why this kidney disorder is called Carp-dropsy complex.

Similar to Renal dropsy, there is renal damage, along with an enlargement of the fish's eye (exophthalmos). The treatment is usually unsuccessful and death occurs within six months.

3. Proliferative kidney disease is caused by the PKD parasite, and has become the most important kidney and urinary tract disorder to affect the fish industry. It usually occurs in rainbow trout and other fishes belonging to the salmon family. Proliferative kidney disease infects young fish, generally during summers when the temperatures rise above 12 degrees Celsius.

Fish with the disease will display sluggishness, bulging of the eye (exophthalmos), renal dropsy, accumulation of fluids in the abdomen, and swelling of the lateral side of the body. Unfortunately, treatment for Proliferative kidney disease is not usually successful.

## HEMATOLOGIC DISORDER

Hematology can be a useful tool for monitoring health status, detecting illness, and following the progress of disease and response to therapy. Despite advances in fish medicine in recent years, interpretation of fish hematology often is hampered by a lack of meaningful reference values and the bewildering diversity of fish species. A multitude of intrinsic and extrinsic factors cause normal and abnormal variation in hematologic data.

### Eye Disorders In Fishes

Eye disorders in fishes can be due to disease, infections, or injury.

### Symptoms and Types

These disorders can cause the eye(s) of the affected fish to display any of the following symptoms:

- Swelling
- Enlargement (giving an appearance of a popping eye)
- Blood in the eye
- Ulceration
- Disfigurement

- Parasites within the eye
- Abnormality around the eye

A fish's eye is usually examined with a penlight or a flashlight. These are used to ascertain whether the problem is within the eye or in the area surrounding it.

Eye injuries usually occur during shipping and handling, especially if the fish is struggling. Blood in the eye, however, is generally due to infection or injury.

## Treatment

There are many common eye disorders affecting the fishes. The three main eye disorders in fishes are:

1. Gas Bubble Disease: This eye disorder is recognized by the tiny gas bubbles found in the cornea - the thin, transparent tissue covering the eye. The fish may also form tiny bubbles in the gills or fins. Usually a gill biopsy is needed to confirm the gas bubble disease. The veterinarian will then recommend appropriate treatment for the fish.
2. Cataracts: Fishes can also suffer from cataracts, which is a common eye disorder causing the eye lens to become opaque. Cataracts can be due to nutritional imbalance, parasitic infection, and other genetic or unknown factors. Unfortunately, there is usually no treatment for cataracts.
3. Eye flukes: This is a type of parasitic infection, which is usually seen in fishes found in the wild. An infected fish will have enlarged and cloudy eyes, occasionally with tiny worms also being found in the eye. The fish will generally become blind in the infected eye, and it may develop a cataract, too. Unfortunately, there is no treatment for eye flukes.

# PARASITIC DIGESTIVE DISORDERS IN FISH

Most digestive disorders in fishes are caused by parasite infections. However, not all parasites cause problems for fishes -- some live in a symbiotic relationship with the fishes.

## Symptoms and Types

The symptoms can vary depending on the parasite causing the digestive disorders, but usually include weight loss, lethargy and loss of appetite. Young fishes are especially susceptible to digestive disorders and can die before any symptoms arise. Such parasites which cause these type digestive disorders include:

- Protozoan parasites (e.g., Spiionucleus, Hexamit, and Cryptobia)
- Worm parasites (e.g., tapeworms)

Protozoan parasites Spiionucleus and Hexamita infect the intestines of cichlids, bettas, gouramis, and other aquarium fishes. Apart from the symptoms listed above, fishes with these two parasites produce stool that is white and stringy. Cryptobia, another protozoan parasite, infects the stomach of African cichlids.

## Causes

Various living conditions which can cause digestive disorders in fishes include:

- Overcrowding of the aquarium, tank or fishpond
- Shipping methods
- Handling methods
- Infected food
- Stressful conditions

Anti-parasitic medicines can be used to clear out a tapeworm, Spiionucleus, and Hexamita infections. However, there is no treatment for a Cryptobia infection. Those fishes infected with this parasite eventually stop eating and then die. Anti-parasitic medicines can be used to clear out a tapeworm, Spiionucleus, and Hexamita infections. However, there is no treatment for a Cryptobia infection. Those fishes infected with this parasite eventually stop eating and then die.

# ENVIRONMENTAL GILL DISORDERS IN FISH

Gills are special organs which allow fishes to breathe underwater. However, if a fish's environment is not well maintained, it can develop gill disorders. Of these, the three main disorders are gas bubble disease, carbon dioxide toxicity, and hydrogen sulfide toxicity.

1. Gas bubble disease usually occurs in cold water systems. When the water in the tank, aquarium or fishpond has an abnormal amount of dissolved gases (i.e., nitrogen, argon, carbon dioxide), fishes can develop gas bubble disease. This occurs when the water is heated up too quickly or because of a faulty pump - pulling air in with the water - in aquariums or tanks; it can also occur if there is heavy algal growth in ponds.

Fishes affected with this disease develop tiny gas bubbles in their eyes, fins, and gills. It can then be treated by blowing out excess gases out of the water through forceful aeration - stirring up the water - and by fixing any faulty equipment.

2. Carbon dioxide toxicity occurs when the carbon dioxide concentration in water is more than 20 mg per liter. The pH of the water becomes acidic, and thus toxic for fishes.

Fishes with carbon dioxide toxicity are unresponsive to stimulus and lethargic. The treatment involves forceful aeration - stirring up the water - to blow out excess carbon dioxide into the atmosphere and increase the water's pH level.

3. Hydrogen sulfide toxicity can be fatal for fishes. Hydrogen sulfide (H2S) is a gas that forms in aquariums or ponds, when certain bacteria feed on organic debris in the areas of water that are low or depleted in oxygen. In large amounts, H2S is toxic and is identified by a strong sulfurous smell coming from the water.

Fishes with long-term exposure will become thin and sickly, and develop extensive gill damage. Treatment for this particular toxicity involves keeping the water clean of all debris and aerating the water.

## Prevention

Regularly test the water for pH and gas levels to prevent environmental gill disorders. Heating water slowly avoids trapping an excess amount of gases in the water, as does keeping the water clean and well maintained.

## Swim Bladder Disorder

In a fish's world, being unable to right ones-self may be a sign of a swim bladder problem commonly referred to as "Swim Bladder Disorder", "Swim Bladder Disease" or "SBD" for short.

In simple terms, the swim bladder is a gas filled sac located in the betta's body, posteriorly (toward the tail end). You can usually see it bump out slightly on most Bettas. The swim bladder works very similarly to a SCUBA diver's BCD (buoyancy control device). When a diver wants to be more buoyant, he fills his vest with air and releases air when he wants to descend. In fish, it's a little more complicated but you get the basic idea.

SBD is probably caused by overfeeding most often in Bettas. Feeding too much food or foods with a very low moisture content (Betta pellets) may cause pressure in the abdomen cutting off a Betta's ability to regulate the gasses in the swim bladder. This becomes apparent when the fish can no longer swim upright in the water.

The excess food in the stomach may also show itself in the form of bloating around the abdomen like you described. Usually, the easiest fix for this is to fast your Betta for a couple of days until the food passes and the fish is able to regulate the swim bladder again. Once he is back to normal, closely monitor his feedings, making sure only to feed a few pellets a couple times a day.

It's also recommended when feeding dry foods that you soak them in a cup full of tank water for about 10 minutes prior to feeding to allow them to swell to their true size before entering the Betta's digestive track. Some Bettas are prone to bloating and constipation and may benefit from a diet of mostly live or frozen foods of which a variety are available commercially.

A bacterial infection of the swim bladder may also cause the symptoms described. If you don't find thatthe Betta is back to normal after a day or two of fasting it may be necessary to treat the fish for the infection using a wide spectrum antibiotic available at most local fish stores. Because antibiotics can cause additional stress to fish, it is recommended to only use them if an infection is strongly suspected. Other signs of infection include lethargy, color loss, loss of appetite, redness under the scales (septicemia) and other secondary infections.

On some occasions Bettas won't fully heal from their swim bladder disorder and may continue to experience problems righting itself. As long as the fish is able to draw air from the surface and can still eat there is no reason that it can't continue to live out a full life. Euthanization is rarely needed. You can make your Betta more comfortable by keeping him in a tank away from fin nipping fish or fast swimming fish that may agitate him. Adding some floating plants or other decor may help him rest near the surface when he wishes.

Just a note…

• Antibiotic resistance is a problem facing aquarists all over the world. This occurs when bacteria evolve to make newer, stronger strains that cannot be combated by available antibiotics. To avoid creating tougher strains of bacteria be sure to carefully follow the directions on the medicine package and to never stop or change medications part way through treatment unless your betta is having a dangerous reaction to them. As with any disease, remove your sick fish from the community tank and isolate him for treatment in a hospital tank. Never medicate healthy fish.

## SALT WATER FISH DISEASES

The most effective way to deal with diseases is to prevent it by minimizing stress."A varied diet, stable water temperature, good water parameters, daily observance, and responsible stocking are a few ways to ensure healthy fish, free of disease." - Drs Foster and Smith But even if you manage to deal with it, unfortunately it will still strike one or more of your saltwater fish. Fish will get sick.

Pathogens that cause diseases are in and around your fish's natural setting and in your aquarium. They may be bacterial, viral, fungal or parasitic.

So when the fish's skin or tissue gets damaged, whether from shipping, netting or any other circumstances, bacterial and fungal pathogens seize the opportunity to infect the fish.

Poor water quality also weaken the fish's immune system. This could be another cause for fish diseases. Check this page to learn and understand the saltwater aquarium water parameters. Or visit this page to learn more what to do if you're experiencing poor water quality condition in your tank.

The bad thing is that there's not a lot of treatment available for us aquarists and there is no guarantee that our saltwater fish will be saved.

We will cover topics that are most common in the home aquarium - how you'll be able to recognize these fish diseases and treatments available.

- Bacterial Infections
    - Fin Rot
    - Fish tuberculosis, wasting disease
    - Vibriosis, ulcer disease
- Viral disease
    - Cauliflower or Lymphocystis
- Fungal disease
    - Ichthyophonus disease, whirling disease
    - Exophiala disease
- Parasitic Infestations
    - Marine velvet

- Marine white spot, cryptocaryoniasis, marine ich
- Uronema disease
- Tang turbellarian disease, black spot
- Tremode infestation
- Crustacean infestation

- Other health problems
  - Head and lateral line errosion
  - Poisoning

## BACTERIAL INFECTIONS

There are good and bad bacteria. And all bacteria are microscopic one-celled organisms are capable of reproduction at a very fast rate.

Remember that some bacteria are beneficial to the aquarium in the nitrogen cycle? But some can cause infections.

Here are our list of bacterial infection fish diseases and treatments.

- Fin Rot
  - o Causes

Aeromonas, Pseudomonas, Vibrio bacteria

- o Symptoms

This bacterial infection causes erosion or rotting of the fins and the fin rays. The base of the fins reddens. In an advance stage, it spreads to the skin causing bleeding to the gills.

- o Treatment

You can remove uneaten food, do a partial water change, and change the activated carbon in your filter.

The antibiotics furanace, augmentin, and ciprofloxin may be effective.

Recommended treatments: Melafix, Furanase, Furan-2, Maracyn, Maracyn-Two and Maracyn Plus.

- Fish tuberculosis, wasting disease
  - o Causes

Mycobacterium bacteria

- o Symptoms

There's really no external sign for this kind. If your fish looks fine from the outside appearance and looks healthy, it will be hard to recognize if it is internally infected.

Fish that are infected though may live a year or more. Skin lesions, emaciation, labored breathing, scale loss, frayed fins and loss of appetite are all signs of this infection. Unfortunately, by the time these symptoms are recognized, it is probably too late to save the fish.

- o Treatment

This bacteria are transmitted orally through raw infected fish flesh, detritus, and feces of the infected fish. They can also infect skin wounds and lesions.

The best treatment is prevention. Try not to feed raw fish and shellfish to your saltwater fish. Antibiotics, including kanamycin, erythromycin, and streptomycin, can be used.

If the saltwater aquarium is heavily infected with this type, it must be sterilized and the water should be discarded.

- Vibriosis, ulcer disease
  - o Causes

  Vibriobacteria
  - o Symptoms

  Variety of symptoms are associated with this disease. They can include lethargy, darkening of color, anemia, ulcers on the skin and lower jaw, bleeding of the gills, skin and intestinal track, clouded eyes, loose scales, pale gills and sudden death.
  - o Treatment

  This bacteria commonly go straight to the intestinal tracts of a healthy fish. They become dangerous only when stress allows infection.

  Poor water quality, crowding, excessive handling, and copper treatments are common causes of stress in saltwater aquarium fish. Immersion treatments with antibiotic compounds have met with some success.

## Viral disease

Viruses are microscopic organisms that thrive by invading the cells of their hosts. In most cases, there are no treatments for the few viral diseases of the saltwater aquarium.

Here are our list of viral fish diseases and treatments.

- Cauliflower or Lymphocystis
  - o Causes

  Cystivirusvirus
  - o Symptoms

  Fin and body lesions are raised, whitish, warty and have a lumpy texture like a cauliflower. It may take three to four weeks for these lesions to reach their full size.

  The infection is not really fatal, but it can be transmitted to other fish in the saltwater aquarium.
  - o Treatment

  The only treatment is to isolate the fish immediately and let the fish's immune system deal with the infection. This may take as long as several months. Other hobbyist scrape the lesion off the fish.

## Fungal disease

Fungi are plan-like organisms. Some are parasitic on fishes.

- Ichthyophonus disease, whirling disease
  - o Causes

  Ichthyophonus fungus

o Symptoms

These fungi goes to the internal organs, infecting the kidney, heart, spleen and liver.

Signs include emaciation, spinal curvature, darkening or paling of the skin, roughening of the skin, fin erosion, and skin ulcers. Erratic swimming behavior can be a symptom too.

o Treatment

This parasitic organism has a complex life cycle. The fungal cysts are usually ingested by the fish, where thy burst and enter the bloodstream, infecting the internal organs.

Fishes that has this kind of disease typically die up to two months after being infected. Treatment is hard because of the internal nature of this disease.

The infected fish should be removed right away from the aquarium to prevent other fishes from becoming infected.

Recommended treatments: Rid-Ich+ or Probiotic Marine Formula. For scaless fish: Rid-Ich+ or Ich-Attack. Learn more about this treatments here.

- Exophiala disease

o Causes

Exophialafungus

o Symptoms

Lethargy, disorientation, and abnormal swimming are the signs.

o Treatment

No treatment is known. The best thing to do is to isolate the fish to prevent others fishes from contracting the fungus.

## Parasitic Infestations

- Marine velvet

o Causes

The dinoflagellate protozoan Amyloodinium ocellatum

o Symptoms

You can usually see the infection on the gills then it spreads to the skin. It becomes dull, patchy and velvet-like. White spots are visible on section of intact skin.

The fish's behavior may include fasting, gasping, scratching against objects, and sluggishness. Death can occur in as little as 12 hours if infected. Tangs, clownfish, and angelfish are especially susceptible.

o Treatment

The organism has three stages to its life cycle and only one of them is parasitic. No effective treatment is known, but copper-based products and some antibiotics are effective.

A freshwater dip sometimes digs out the parasites from the host but does not kill them. Treatments are often dragged and the entire aquarium must be treated to fully destroy the infestation.

- Marine white spot, cryptocaryoniasis, marine ich
  - o Causes

  The ciliate protozoan Cryptocaryon irritans
  - o Symptoms

Early signs are fasting, cloudy eyes, troubled breathing, excess skin mucus and pale skin. White spots appear on the skin, gills, eyes and death usually follows within a few days.

- o Treatment

White spot organisms are hard to control. The freshwater dip may be effective in killing parasite on the fish but does little to the aquarium. Prolonged immersion in copper products is the most effective treatment.

- Uronema disease
  - o Causes

  The ciliate protozoan Uronema marinum
  - o Symptoms

External ulcers, muscle and skin bleeding, lethargic behavior, cloudy eyes, and sloughing of the skin are signs. They usually die right away if infected because of the impaired circulation of the gills.

- o Treatment

Freshwater bath followed by prolonged immersion in a formalin bath a day later is effective.

- Tang turbellarian disease, black spot
  - o Causes

  Paravortex flatworms
  - o Symptoms

Even if the name implies that only tangs are infected, this is not the case. Many can be infected by this flatworm.

The early phase, the organism would look like numerous dark spots over the fins, gills, and body. Other signs include fastening, listlessness, paling or whitish skin, and scratching against objects.

- o Treatment

As most parasitic infestations, crowding allows the disease to spread to other fish in the aquarium. Freshwater bath, triclorfon, formalin or praziquantel immersion can be effective.

- Tremode infestation
  - o Causes

  Monogenetic trematode worms
- o Symptoms

Many species of this worms are too small to see without a microscope. They normally infect gills, eyes, skin, mouth and anal opening.

The infected fishes usually rub themselves against objects in the aquarium, trying to remove these parasites. These often causes damage that leads to bacterial infections.

o Treatment

Immersion in fresh water, mebendazole, praziquantel or trichlorion is effective against trematodes.

- Crustacean infestation

o Causes

Copepod, isopod, and argulid crustaceans

o Symptoms

Most of these tiny, crab-like organisms are visible to the naked eyes. Copepods remain fixed in the same position while isopods are argulids move over the surface of the fish, causing tissue damage.

Fishes that are infected swim erratically, rub against objects in the aquarium and jump. Bacteria infect lesions.

o Treatment

Remove the ones that are infected immediately. Remove the decorations as well and dry them to kill the egg masses or immerse them in 2 percent bleach solution for two hours.

Treat the infected fish with fresh water, trichlorfon or malathion baths.

## Other health problems

Just when you thought there could be no more, here are a couple more that you need to be aware of.

- Head and lateral line errosion

o Causes

Poor water quality, nutrient deficiency, possibly a parasite

o Symptoms

Like the freshwater disease called hole-in-the-head, holes develop and enlarge in the sensory pits of the head and down the lateral line on the body.

The disease progress slowly and the fish does not seem to behave differently. Advanced stages can lead to bacterial infection and death.

o Treatment

There are no specific treatment, but some recommended the use of the freshwater antibiotic flagyl. Check your water quality and make the necessary adjustments. Make sure that you are also meeting the fish's nutrional needs. Get a variety of food for their diet and add vitamin supplements to their food.

- Poisoning

o Causes

Multiple causes including buildup of nitrogenous compounds (ammonia, nitrite), household chemicals (smoke, cleaners, fumes) and tap water constituents (heavy metals, chloramines)

o Symptoms

Low levels of toxins in the water stress the fish that can lower their resistance.

Higher levels causes abnormal behavior such as darting movements, jumping, and gasping at the surface of your aquarium water.

o Treatment

Be sure that activated carbon is used to remove toxins. Do a water change, about 20-40% of water should be replaced.

## EPIZOOTIC ULCERATIVE SYNDROME IN FISH

Epizootic ulcerative syndrome (EUS) is a fungal disease of freshwater and brackish fish affecting more than 100 fish species. It is caused by the fungal species, Aphanomyces invadans. The organism requires a specific combination of factors in order to germinate within the dermis of the fish. The disease causes lesions in both the skin and visceral organs.

### Distribution

Disease is present in parts of the Asia-Pacific region and Australia. A similar disease has also been reported in the eastern USA.

### Signalment

Over 100 species have been confirmed to be affected by EUS.

EUS occurs commonly during periods of low temperature and heavy rainfall in tropical and sub-tropical waters. These conditions favour sporulation and cold temperatures delay the inflammatory response of the fish to infection.

Some species have been shown to be resistant including Chinese carps, milkfish and tilapia.

### Clinical Signs

In early disease, red spots or small haemorrhagic lesions are found on the surface of the fish.

These progress to ulcers and eventually large necrotic erosions. Fungal mycelium is often visible on the surface of ulcers.

Death then follows rapidly due to visceral granulomata, septicaemia and failure of osmoregulatory balance.

### Diagnosis

Positive diagnosis is made by analysis of histological sections demonstrating mycotic granulomas and isolation of the causal fungus.

Histology of early lesions reveals acute spongiosis and epithelial cell loss. Degenerative changes progress through the dermis with hyperaemia, haemorrhages and inflammatory infiltration. In advanced stages, sarcolysis is also obvious. Fungal hyphae are enclosed by a well defined epithelioid cell layer and mycotic granulation spreads through the infected muscle and internal organs.

Muscle fibres eventually disappear altogether and are replaced by fibrosis, inflammatory cells and new blood vessels. These distinct features of EUS ulcers make histological analysis enough for a definitive diagnosis.

Squash preparations of skeletal muscle from beneath an ulcer will also demonstrate septate fungal hyphae for a rapid provisional diagnosis.

Moderate, pale, raised lesions are best for fungal isolation. Muscle should be exposed in a sterile manner by removing scales and searing the skin with a red hot spatula before dissecting with a sterile scalpel. 2mm samples should be placed in a petri dish containing Czapex Dox agar with penicillin G and oxolinic acid. They should be incubated at room temperature and examined daily so that emerging hyphal tips can be transferred onto fresh plates to produce contaminant free cultures.

The fungus can then be identified by inducing sporogenesis and demonstrating its typical asexual characteristics.

Injection of spore containing innoculum into susceptible fish at 20?C and demonstration of histological growth after 7 days and granulomas in muscle after 14 days is also diagnostic.

## Treatment

Keeping diseased fish in good quality, clean water may allow recovery, but only if lesions are not too extensive and dark scars are often left behind on healing.

There is no effective treatment for advanced disease.

## Control

Quarantine and health certification practices for movement of live fish between countries is the method of preventing spread of EUS to currently free areas.

In endemic areas, eradication, exclusion, management, surveillance and treatment are all required to gain control.

# EPIZOOTIC ULCERATIVE SYNDROME IN INDIA

The dreadful fish disease, Epizootic Ulcerative Syndrome (EUS), has been causing large-scale mortality among the freshwater fishes since 1988, initially affecting four species of fishes widely. The study revealed the fluctuation in intensity of disease in relation to species effect. Large hemorrhage cutaneous ulcers, epidermal degeneration and necrosis followed by sloughing of scales are the principal symptoms of EUS.

Low total alkalinity could be a pre-disposing ‘stress’ factor. Sick fishes show low haemoglobin and polymorph, but high ESR and lymphocytes. Communicative nature of EUS revealed a variation in time-gap between fish and infection in different species. Inoculation of microbes into the test animals did not reveal any sign of ulceration for two years.

Bacterial culture revealed occurrence of haemolytic E. Coli, Aeromonas hydrophila, Pseudomonas aeruginosa, Klebsiella sp., and Staphylococcus epidermitis in the surface lesions as well as in the gut, liver, gills, heart, kidney and gonads of sick fishes, all of which have been found to be sensitive to Chloramphenicol, Septran, Gentamycin, etc.

Fungal isolation revealed occurrence of Aphanomyces sp with concomitant occurrence of the same fungal genus in histological sections of EUS affected fishes. Histopathological studies revealed focal areas of fatty degeneration of hepatocytes surrounding the portal triads in the liver.

The preliminary histochemical studies with regard to interruptions in glycogen synthesis and blockade of respiratory pathways are being conducted. Similarly, preliminary enzymological studies are being conducted with regard to the amount of alkaline phosphatase, SGOT, SGPT and LDH.

Inoculation of 10% homogenate from EUS affected Clarias batrachus into 80% confluent monolayer of BF 2 fish cell line in Leibovitz L-15 medium revealed progressive CPE at a dilution of 10 which was passable in subsequent cultures; indicating the isolation of the virus. Electron microscopic studies with the ultra-thin sections of still-occurring EUS - affected fish tissues, revealed the presence of virus-like particles (inclusion bodies); and further studies in this regard are being conducted.

## INTRODUCTION

Fish is a dependable source of animal protein in developing countries like India. Large-scale mortality occurs among the fresh water fishes often due to environmental stress followed by pathogenic attacks and parasitic afflictions (Epizootic Ulcerative Syndrome, EUS) causing a tremendous loss to the nation.

Such an epizootic, hitherto unknown, assumed epidemic dimension and has been sweeping a big part of the globe starting from Australia and tropical South-America during early seventies and reaching South - East and south Asia down the years; and in these regions, this epizootic is persisting even today, posing a serious threat to the existence of freshwater fish.

Following reports of Mycotic granulomatosis (MG) in Japan (1972), the earliest report of EUS in Papua New Guinea (1974), most of south-east and south-Asian countries like Indonesia (1980), Malaysia (1979-83), Thailand (1981-85), Kampuchea and Lao PDR (1984), Myanmar (1984-85), Sri-Lanka (1987), Bangladesh (1988), and the Barak valley region of Assam, India (1988). Since then, it has been spreading rapidly causing large-scale mortality among freshwater fishes.

### Major Species Affected And Pattern Of Spread Of Eus

A systematic study undertaken since 1988 in India, revealed wide scale attack among four species of fishes, viz, Channa punctatus, Macrognathus aral, Mystus vittatus and Puntius conchonius. Other species affected by EUS during the same period included Aorichthys aor, Amblyphayngodon mola, Catla, Cirrhinus mrigala, Heteropneustes fossils, Labeo rohita, Lepidocephalus guntea, Notopterus notopterus and Salmostoma bacaiia.

The study revealed EUS attack among different species of fishes (other than those mentioned above) during the subsequent years (Kar and Dey 1990 a,b).

The recent studies (Kar et.al., 1993;1994;1995a,b; 1996;1997;1998a,b,c;1999 a, b,c,2000a, b) indicated that species like Channa striatus, C. Punctatus, Clarias batrachas and Anabas testudineus have been severely affected by EUS in this region since 1995 and the outbreak has been occurring during the period from November to March. This indicates that there is a differential pattern of species susceptibility as EUS has been spreading through different years.

## Symptoms And Gross Pathology

EUS is omogen-ized by the occurrence of large haemorrhagic or necrotic ulcerative lesions on the base of fins and other parts of the body, which later become larger inflamed areas with acute degeneration of epidermal tissues.

## Pattern Of Investigation Of Eus

Sniesko (1974) stated that an overt infectious disease occurs when a susceptible host is exposed to virulent pathogen under stress. Field estimation of temperature, turbidity, pH, Dissolved oxygen, free carbon dioxide and conductivity in water; temperature, pH, conductivity, organic carbon and NPK in soil; aquatic macrophytic biomass (AM) and total planktonic count did not reveal any noticeable differences in the values of these parameters obtained during the disease outbreak period in the similar water bodies.

This could indicate that, there had not been any severe organic pollution of water and soil in these water bodies; and hence organic pollution of water and soil may not be considered as the principal cause of EUS outbreak in this region. However there has always been severe fall in total alkalinity (TA) values in these water bodies prior to EUS outbreak; thereby causing 'stress' condition for fishes in the water and making them susceptible to attack by the pathogens; leading to the outbreak of EUS (Kar and Dey, 1990a).

Estimation of mercury, lead, Arsenic and Cadmium (by atomic adsorption spectrophotometry), did not reveal any appreciable differences in the quantity of these elements in the water and soil samples of the EUS affected and unaffected waterbodies. This had indicated that, absence of trace element contamination, in the outbreak of EUS (Kar and Dey, 1990 b).

# AETIOLOGICAL STUDIES

## Bacteria

Regular isolation of bacterial flora from the surface lesion of EUS affected fishes as well as, from, their muscles, gut, liver, gills, heart, kidney and gonads revealed the occurrence of haemolytic strains of Eschericha coli, Aeromonas hydrophila, Pseudomonas aeruginosa, Staphylococcus epidermitis and Klebsiella sp. All these bacteria have been found to be sensitive to chloramphenicol, septran, streptomycin, gentamycin and norfloxacin (Kar et.al, 1999 a).

## Virus

Absence of haemorrhagic septicaemia (which is so characteristic of Aermonas infection) in all but most of the ulcerated fish (Roberts 1989); non-occurrence of these pathogens in the early stages of the disease (Lilley et al., 1992); and close association of these bacteria with the fish since time immemorial (Kar and Dey 1990 b) suggest that A. hydrophila is unlikely to have any primary role in the pathogenesis of EUS. And, a primary viral aetiology has always been suspected to be associated with the initiation of EUS (FAO 1986; Kar et.al., 1990 a).

As a confirmation for the presence of viruses in the EUS - affected tissues, BF 2 fish cell lines was used by infecting the growing cells with filtrate from omogenized tissue of ulcerated Clarias batrachus. A progressive CPE was noticed even at dilution of 10 of a 10% tissue homogenate. The infection was less in the subsequent cultures indicating the isolation of virus (Kar et.al., 1993). Electron microscopic study of the TCF showed occasional round particles unlike those previously reported (Frerichs et.al., 1986).

# ELECTRON MICROSCOPIC STUDIES

Detailed electron microscopic studies done with ultra thin sections of EUS - affected fish tissues revealed the membrane bound and granular abnormal structures; hepatocytes with vacuolations and virus-like inclusion bodies in the nucleus.

## Fungi

Kar et.al., (1999b) isolated Aspergillus sp from diseased fishes in Assam. Subsequently, Kar et. Al (2000a) also isolated Aphanomyces sp from the EUS affected fish tissues. However, fungi may not be considered as the primary aetiology in view of the fact that, they have always been associated with fishes in water without generally causing any harm.

## Histopathology, Haemotology, Histochemistry And Enzymology

Detailed histopathological (HP) studies (Kar and Upadhyay 1998a) revealed focal areas of increased fibrosis and chronic inflammatory cell infiltration in muscles; focal areas of fatty degeneration of the hepatocytes surrounding the portal triads and occasional infiltration by mononuclear cells in the liver. Preliminary haemotological studies revealed higher DLC and ESR but low Hb content in the blood of EUS affected fishes as compared to the corresponding healthy fishes of the same species (Kar et.al., 1994).

Preliminary histochemical studies (Kar and Das 1999 b) revealed interruption of glycogen synthesis in the liver and blockade of respiratory pathways in the gills. Preliminary enzymological studies (Kar 1999a) in the EUS - affected fishes revealed high values of alkaline phosphatase, SGOT, SGPT and LDH.

### Control Of Eus

Effective control and treatment of EUS is a major problem today, because the primary pathogen has not yet been identified. Control of EUS in large natural water bodies may not be possible. However in view of the fact that TA of water acts as a predisposing 'stress' factor, application of lime in low alkaline water bodies, helps to control the spread of EUS. EUS today is a semi-global problem among the fresh water fishes. Unfortunately, in view of its complex infectious etiology, it is yet to be accurately defined.

## FISH DISEASES IN AQUACULTURE

Therefore, both the aquaculture industry itself, and mandated governmental fish health management programs, maintain a variety of measures to control and minimize the effect of infectious diseases in culture fish populations in fish farms and aquaculture sites. Nevertheless, diseases can and do occur in these populations. This paper addresses the concern that diseases can spread from aquaculture sites to wild fish with resulting effects on populations of wild fish.

## DYNAMICS OF FISH DISEASES

The dynamics of the occurrence, severity and spread of diseases within and between fish populations are similar to those associated with diseases in human and terrestrial animal populations. However, one additional component in fish populations is the water environment which may facilitate the spread of disease-causing agents.

Diseases are transmitted from one individual to another in one of two ways: vertical transmission or horizontal transm-ission. In vertical transmission, pathogens are transmitted from one or both parents to offspring through sex cell which for offspring. Horizontal transmission involves the spread of the pathogen from one individual to another through direct contact, air, or water.

The development and severity of a disease following exposure to a pathogen involves a complex web of variables such as: the virulence (ability to cause disease) of the pathogen; the immune, genetic and physiological condition of the host; stress; and population density. Different strains of the same pathogen may vary considerably in their ability to cause disease. That is, some strains can infect a host without causing a clinical disease. Also, individual animals differ in susceptibility to disease.

Some individuals may be more resistant to the disease because of their genetic composition or as a result of previous exposure and development of immunity. Conversely, individuals may be more susceptible to a disease because of poor nutritional status, stress or any of a number of other factors. Population density is a particularly important factor in the spread of fish diseases.

Diseases can spread more easily within dense populations simply because of increased opportunities for association of infected fish with uninfected fish. The actual development of a disease and the relevant severity of that disease within a fish population is influenced by a complex interaction of these variables associated with the pathogen, the host and the environment.

## Diseases in Aquaculture Populations

Infectious diseases pose one of the most significant threats to successful aquaculture. The maintenance of large numbers of fish crowded together in a small area provides an environment conducive for the development and spread of infectious diseases. In this crowded, relatively unnatural environment, fish are stressed and more susceptible to disease. Moreover, the water environment, and limited water flow, facilitates the spread of pathogens within crowded populations.

## Control of Diseases in Aquaculture Populations

In their self-interest, the aquaculture industry in Maine and elsewhere in North America and Europe places a high priority on the prevention and control of infectious diseases. Unlike treating human or other animal diseases, few drugs are available for treating diseases in fish. Because of environmental concerns, only two drugs are approved in the United States for use in fish populations. Control of diseases in aquaculture and fish farms relies on a combination of good management practices, use of the few approved and commercially available drugs and vaccines, and prevention of infection.

There are two general categories of fish diseases: diseases caused by pathogens that are indigenous (native) to the local environment and exotic pathogens that normally are not found in that geographical area. Control of diseases caused by indigenous pathogens is attempted through good management practices and the use of approved drugs and/or vaccines.

Governmental agencies in the Northeast and elsewhere in the United States and other countries have established fish health management programs that mandate periodic testing of aquaculture populations and certification that these populations are free of specific exotic or high risk pathogens. Importation to the area or movement within the area is restricted to fish and fish eggs that have been certified to be free of these pathogens.

It is important to note that such fish must be "free of the pathogen,"not simply without overt disease symptoms. Such disease control programs depend on the availability of sensitive, rapid and cost-effective diagnostic assays. The development of such diagnostic assays is improving rapidly, particularly as a result of current research adapting new developments in molecular biology and biotechnology to fish disease diagnostics.

## Diseases in Wild Fish

The fish pathogens found in aquaculture populations also occur in wild fish. Except for "schooling" fish, the population densities of wild fish are comparatively low and the concentration of pathogens from an infected fish is rapidly diluted in the water environment. Although there are a few historical examples of disease outbreaks that significantly reduced certain populations of wild fish species, it generally is thought that diseases do not play a major role in influencing wild fish populations. However, the exact role, if any, of diseases in regulating populations of wild fish is largely unknown. The introduction of an "exotic" pathogen probably poses the biggest threat to wild fish populations.

### Disease Implications of Interactions of Aquaculture and Wild Fish Populations

Fish in aquaculture populations, particularly those in ocean net pens, can be exposed to pathogens from wild fish in the area. Because of the conditions of the aquaculture environment, these cultured fish populations may be more at risk of developing disease than the wild fish. There is much scientific documentation of the introduction of pathogens and resulting disease outbreaks into cultured fish populations from naturally infected wild fish.

Transmission of diseases also could occur in the opposite direction (i.e., from aquaculture populations to wild populations). In this regard, the greatest concern would be exposure of wild fish to larger than normal concentrations of indigenous pathogens or exotic pathogens from aquaculture sites. However, the fish health management programs mandated in Maine and Maritime Canada significantly reduce the latter threat of introduction of exotic pathogens into wild fish populations. While the spread of diseases from aquaculture populations to wild fish is possible, there is little scientific evidence of such transmissions or, at least, harmful effects of any such transmissions.

### Human Health Implications of Fish Diseases

The vast majority of fish diseases are restricted to fish and pose no risk to handlers or consumers.

## THE ROLE OF STRESS IN FISH DISEASE

### Stress

Physiological stress and physical injury are the primary contributing factors of fish disease and mortality in aquaculture. Stress is defined as physical or chemical factors that cause bodily reactions that may contribute to disease and death. Many potential fish disease pathogens are continually present in the water, soil, air, or fish.

In nature fish are often resistant to these pathogens, and they are able to seek the best living conditions available. Food fish reared under commercial aquaculture conditions are confined to the production unit and are weakened by stress conditions including:

- increased fish density and poor water quality (i.e., low dissolved oxygen, undesirable temperature or pH, increased levels of carbon dioxide, ammonia, nitrite, hydrogen sulfide, organic matter in the water);
- injury during handling (i.e., capture, sorting, shipping);
- inadequate nutrition;
- and poor sanitation.

These conditions can result in decreased resistance by the fish, resulting in the spread of disease and parasite infestation. Stress and injury initially trigger an alarm reaction (fight or flight response), which results in a series of changes within the fish. A blood sugar increase occurs in response to hormone secretion from the adrenal gland as liver glycogen is metabolized.

This produces a burst of energy which prepares the animal for an emergency situation. In addition, the inflammatory response, a defense used by fish against invading disease organisms, is suppressed by hormones released from the adrenal gland. Water balance in the fish (osmoregulation) is disrupted due to changes in the metabolism of minerals.

Under these circumstances, freshwater fish absorb excessive amounts of water from the environment (over-hydrate); saltwater fish lose water to the environment (dehydrate), This disruption increases energy requirements for osmoregulation. Respiration increases, blood pressure increases, and reserve red blood cells are released into the blood stream.

Fish are able to adapt to stress for a period of time; they may look and act normal. However, energy reserves are eventually depleted and hormone imbalance occurs, suppressing their immune system and increasing their susceptibility to infectious diseases.

## DEFENSE AGAINST INFECTION

### Mucus

Mucus (slime layer) is the first physical barrier that inhibits entry of disease organisms from the environment into the fish. It is also a chemical barrier, containing enzymes and antibodies which can kill invading disease organisms. Mucus also lubricates the fish, aiding their movement through water, and is important for osmoregulation.

Injury as a result of handling (i.e., capture, transport, etc.) and certain chemicals in the water (i.e., poor water quality, disease treatments) remove or damage the mucous layer, reducing its effectiveness as a barrier against infection at a time when it is needed most. This damage decreases the chemical protection of the slime layer and also results in excessive uptake of water by freshwater fish and dehydration by saltwater fish. Decreased lubrication causes the fish to expend more energy to swim at a time when its energy reserves are already depleted.

## Scales and skin

Scales and skin function as a physical barrier which protects the fish. These are injured most commonly by handling, rough surfaces of tanks or cages, and by fighting caused by overcrowding or reproductive behavior. Parasite infestations can also result in damage to gills, skin, fins, and loss of scales.

Damage to scales and skin of the fish can increase the susceptibility to infection. It also causes excessive uptake of water by freshwater fish or loss of water from marine species (osmotic stress). Fish which are heavily parasitized may die from bacterial infections which gained initial entrance to the fish's body through damaged areas in the skin.

## Inflammation

Inflammation is a natural immune response by the cells to a foreign protein, such as bacterium, virus, parasite, fungus, or toxin. Inflammation is characterized by swelling, redness, and loss of function. It is a protective response, an attempt by the body to wall off and destroy the invader.

Any stress causes hormonal changes which decrease the effectiveness of the inflammatory response. Temperature stress, particularly cold temperatures, can completely halt the activity of the immune system, eliminating this defense against invading disease organisms.

Excessively high temperatures are also extremely detrimental to the fish's ability to withstand infections. High water temperature may favor rapid population growth of some pathogens. High temperature also reduces the ability of the water to hold oxygen and increases the metabolic rate and resulting oxygen demand of the fish.

## Antibodies

Unlike inflammation and other nonspecific forms of protection, antibodies are compounds formed by the body to fight specific foreign proteins or organisms. The first exposure results in the formation of antibodies by the fish which will help protect it from future infection by the same organism.

Exposure to sublethal concentrations of pathogens is important for fish to develop a competent immune system. Animals raised in a sterile environment will have little protection from disease. Young animals may not have as effective an immune response as older animals and therefore, may be more susceptible to pathogens in the environment.

Stress impairs the production and release of antibodies. Temperature stress, particularly rapid changes in temperature, severely limits the fish's ability to release antibodies, giving the invader time to reproduce and overwhelm the fish. Prolonged stress reduces the effectiveness of the immune system, increasing the opportunities for disease-causing organisms.

# DISEASE PREVENTION

Numerous books and articles have been written on the diagnosis and treatment of specific fish diseases; however, prevention through good management practices is the best control measure to minimize disease problems and fish kills. Good management involves maintaining good water quality, preventing injury and stress during handling providing good nutrition, and using sanitation procedures, The following are management practices that help prevent stress and the resulting fish kills.

## Water quality

1. Do not exceed carrying capacity of fish in ponds and tanks.
2. Monitor water quality parameters.
3. Maintain dissolved oxygen levels above 5 mg/L. Sub-optimum levels of dissolved oxygen, while not immediately lethal, may stress fish, resulting in delayed mortality.
4. Prevent the accumulation of organic debris, nitrogenous wastes (ammonia and nitrite), carbon dioxide, and hydrogen sulfide.
5. Maintain appropriate pH, alkalinity, and temperature for the species.

## Handling and transporting

1. Use capture methods that minimize physical injury and stress.
2. When possible, use knitted mesh nets rather than knotted nets to reduce injury and scale loss.
3. Speed and gentleness when handling fish are of utmost importance.
4. Minimize the number of times the fish are lifted from the water, and work as quickly as possible when transferring fish.
5. Harvest, handle, and transport fish at times when fish are least susceptible to stress and infection.
6. Transport and holding tanks should be large enough to allow complete freedom of movement of fish and have no sharp corners or edges that might injure the fish.
7. Maintain optimum water conditions while capturing, hauling, and handling fish.
8. A high level of dissolved oxygen is crucial for rapid recovery of the fish from the struggle of capture and handling.
9. Salt (0.3 to 1.0 percent) maybe used in the transport water to minimize osmotic stress and bacterial infection of freshwater fish.
10. Ice may be added to the water during hauling to prevent an increase in water temperature which reduces the ability of the water to hold oxygen and increases the metabolic rate and resulting oxygen demand of the fish.

## Nutrition

1. Feed a high quality diet that meets the nutritional requirements of the species.

2. Use proper feeding rate (either over-feeding or starving the fish should be avoided).
3. Store feed in a cool dry place to prevent deterioration. If avail able, a freezer is ideal for storing fish feed.

## Sanitation

1. Quarantine all new fish and observe for mortality. Send samples to a diagnostic laboratory to be examined for parasites and evaluated for viral and bacterial disease.
2. Prevent disease-carrying fish from living in hatchery water supply (e.g., reservoir ponds, springs, streams).
3. Remove all dead fish from a production system as soon as they are observed.
4. Dispose of dead fish properly to prevent the spread of disease.
5. Use good sanitation practices resulting in clean equipment, ponds, and tanks.

   Disinfect containers, nets, and equipment to minimize transmission of parasites and disease from one population to another.

## Conclusions

Stress compromises the fish's natural defenses against invading pathogens. When disease outbreaks occur, the underlying stress factors, as well as the disease organism, should be identified. Correcting stress factors should precede or accompany chemical disease treatments.

A disease treatment is only an artificial way of slowing down an infection so that the fish's immune system has time to respond. Any stress which adversely affects the fish will result in an ongoing disease problem. Prevention of disease outbreaks is more cost-effective than treating dying fish.

# 6

# SCIENTIFIC MANAGEMENT OF FISH DISEASES

Fisheries aims at developing and maintaining sea wealth. Fisheries are estimated to provide 16% of the world population's protein and India holds the seventh position in the world as regards to the value of fish exports. Around 8 million people of India are directly or indirectly dependent on the fisheries sector. The rapidly developing fishery industry, provides employment not only to the traditional fishermen, but also to the highly skilled personnel engaged in the scientific breeding and management of fisheries. The scope in the field of fishery science is increasing with the development of brackish water and freshwater aquaculture. Because of the country's expansive natural resources, dynamic entrepreneurial skills, this profession is advantageous to the Indian economy also.

Fisheries science is the academic discipline of managing and understanding fisheries. It is a multidisciplinary science, which draws on the disciplines of oceanography, marine biology, marine conservation, ecology, population dynamics, economics and management to attempt to provide an integrated picture of fisheries. In some cases new disciplines have emerged, as in the case of bioeconomics.

Fisheries, as the name suggests is an organized effort which is related to the capturing of fish including inland and sea or other aquatic species. This industry involves catching, processing, marketing and conservation of fish. Fisheries Science is an academic discipline of managing and understanding fisheries. This is a multidisciplinary subject that includes the biological study of life, habits and breeding of various species of fish. It also involves farming and husbandry of important fishes and aquatic organisms in fresh water, brackish water and any marine environment.

A career in fisheries science can create an aqua culturist, farm managers, exporters, traders, breeders and modern fishermen's, etc. The main objective of fisheries science is to make available the much needed technically competent extension personnel for transfer of technology. It helps a person to assist the research and development; to develop and execute fisheries development plan.

## PROVIDING WATER SOURCES FREE OF PATHOGENS

Underground waters are the most suitable water sources free of pathogens. These sources are limited both for trout farms and hatcheries and for other special fish culture units at present. The surface water from rivers and channels is used as the source of inflow water in most cases.

In these situations, suitable filters can partially reduce the numbers of invasion stages of parasites in inflow water, above all when supplying smaller reservoirs with intensive culture. Bars are usually placed before these filters to separate rough particles. Sand filters are consisted of a set of sedimentation divisions terminated by filter with fibre and sand.

These type of filters catch above all the heavier parasite stages unable to move actively (e.g. spores). Lower efficiency is registered in elimination of moving parasites like e.g. infusorians.

The water from the pond with fish stock is quite unsuitable for these purposes (esp. as the source of inflow water for trout farms, hatcheries and units for early fish fry stages).

Chemical treatment of inflow water is an emergency arrangement with often undesirable parallel affects. Disinfection of the water entering fish culture units by UV radiation is not still an usual way although it can be considered as the simple method how to destroy viruses, bacteria and moulds germs. Since the inflow water from rivers and channels is slightly turbid and contents a number of suspended solids and dissolved compounds, the disinfective efficiency of UV radiation is markedly reduced in these situations.

It is very profitable to supply the individual ponds and/or reservoirs independently, not throughflowly. The water from each pond or reservoir should be drained separately and should not flow into any other. Especially quarantine ponds and other reservoirs can be separated by this way.

## PROTECTION FROM THE TRANSFER OF PATHOGENS

This principle means above all the transfer of pathogens by uncontrolled transport of fish and spawns. The transport of fish with unknown health condition is to be avoided in principle. All transported fish are to be accompanied by veterinary certificate confirming that fish were examined before transporting them, they are healthy and originate from the environment in which no important transfer diseases appear.

The list of these diseases is precisely stated in veterinary instructions. Except of the internal survey for each country also the list of diseases stated in international codex is obligatory for veterinary service. This list is currently specified with the development of diagnostic methods and improvement of knowledge about individual fish diseases. Some viral and bacterial diseases can be transfered also by spawns. Their transport must be completed by the same veterinary certificate like fish transport from this reason.

Fish introduced from other territories must be subjected to quarantine for one year regardless if native or extraneous species. The duration of quarantine can be prolongated e.g. in the case of fish imported from abroad until the period of 3 years.

Prolongated period of quarantine is of special importance especially in spawners predestined for further reproduction of imported species.

The selfsustaining in stock production in individual farms and similar organizations is a significant way of prevention from dissemination of fish diseases. Only fish previously examined, free of diseases and relevantly treated by medicinal baths are to be stocked into ponds and fish culture units. The stocking of fry originating from semi-artificial and artificial spawning not contacted with fish of higher age categories also minimizes the danger of infection.

The prevention from introduction of coarse fish into ponds and fish culture units is the other important arrangement protecting the stock against transfer of pathogens. These fish are above all the source of ectoparasites, dangerous especially in the period of decreased resistence of fish. Except of this they can transfer also some other pathogens which can result in heavy losses in important fish species. Adequate bars and filters can serve for prevention from coarse fish penetration.

The protection of piscivorous birds to step into fish culture units (esp. trout farms) is the prevention limiting the expansion of some fish diseases. Protective nets are used to prevent the birds from running in. The numbers of piscivorous birds are regulated in localities where overpopulated.

Preventive control of snails (Lymnaea sp.) as intermediate hosts of some fish parasites can be performed by biological (introduction of black carp - Myelopharyngodon piceus or 3-years-old tench Tinca tinca), mechanical (placing nets in the inflow), physical (drying and freezing of the bottom) and chemical (application of molluscocides) ways.

Safe and harmless removing of dead fish is a significant way how to prevent from further transfer of fish pathogens. Fresh or slightly decayed dead fish are decontaminated in the nearest veterinary facility. Lower masses of dead fish are to be burnt or burried into deep pits (aprox. 2 m) in distance of at least 20 m from the pond bank. The bottom of this pit and dead fish must be covered by burnt or chlorinated lime. The layer of at least 60 - 80 cm of the soil must cover the content of a pit.

## DISINFECTION OF PONDS, FISH CULTURE UNITS AND EQUIPMENT

Disinfection is of a big importance in prevention and elimination of fish diseases. Preventive disinfection protects the fish stocks against pathogens. Hygiene of environmental conditions for fish is improved by this way. Focal disinfection is performed for control of the focus of dangerous fish disease.

Natural physical phenomena are fully used for disinfection in intensive fish culture due to their ecomical convenience. It concerns the drying and freezing of the pond bottom. The most of pathogens die after perfect drying of the pond bottom when its relative moisture had dropped on 10 - 15 %. The perfect freezing of the wet places and sun radiation (above all by its UV rays) have a very favourable effect in our conditions.

The influence of these natural physical phenomena is exploited by summer drying and winter freezing of water reservoirs (ponds). Summer drying is a radical, long-term intervention during which all pathogens are controlled due to the perfect drying of the pond bottom. The aim of winter drying is to destroy the pathogens by freeze. It safely leads to destruction of leeches (Piscicola geometra), fish lice (Argulus sp.), predatory larvae of water insects, eggs and spores of parasites and also other pathogens. Employing the natural ways for disinfection has an disadvantage in usually long-term duration (a number of months up to one year).

Chemical disinfection is an effective way of prevention from and/or suppressing of fish diseases. Usually accessible disinfective preparations are used in fish culture (e.g. burnt lime, chlorinated lime, nitrogen lime, natrium hydroxide, potassium permanganate, formaldehyde, Chloramine, Chlorseptol, Jodonal etc.). Burnt lime is mostly employed for disinfection of the bottom of ponds and reservoirs in the dose of 2.5 - 3 t.ha-1, or chlorinated lime in the dose of 0.5 - 0.6 t.ha-1. In case of myxosporoses, nitrogen lime (5 t.ha-1, or 0.5 kg.m-2) is to be applied.

Immediately after fishing out the pond, the disinfection of fishing pit, pond ditches and muddy wet places is performed on large ponds where the whole-surface bottom disinfection is not possible. 5% water solution of formaldehyde, chlorinated lime (200 - 400 mg.1-1), 0.5 % water solution of natrium hydroxide, Chloramine and chlorseptol (30 g.1-1) or other disinfectants can be used for treatment of concrete channels, troughs and other arrangements employed for fish culture.

The same disinfectants and concentrations are to be used for the treatment of the equipment. Potassium permanganate (5 g.l-1), Jodonal (2.8 - 4.5 ml.l-1) and other disinfectants can be also employed for these purposes.

## OPTIMALIZATION OF ENVIRONMENTAL CONDITIONS

The optimalization of natural environmental conditions is the main pre-condition how to ensure the good health condition of stock during the rearing period. The following principles must be ensured:

- optimal water quality without stressing physico-chemical effects. Keeping the oxygen concentration on optimal level and protection against water pollution are of special importance,
- maximum development of natural food resources by the adequate interventions, feeding fish by supplementary feed mixtures in sufficient amount and quality (the attention should be paid on the quality of individual feed components and biofactors), basic preventive arrangements protecting the early developmental stages and young fish from bacteria and protozoans, including sufficient amount of natural food of appropriate size and species composition,
- responsible establishment of maximum stocking density. Inadequately high stocking density results in stress behavior, worsened condition and resistence, and makes the expansion of diseases easier. The stocking density is of special

importance in trout farming and fish culture in special intensive units (but also in ponds),

- prevention from stress situations evoked by other factors, above all manipulation during fishing out, transport and long-term storage.

## REGULAR CONTROL OF HEALTH CONDITION AND PREVENTIVE TREATMENT OF FISH

Preventive control of health condition is to be carried out in hatcheries and early fry rearing units twice a week, and in highly productive intensificated ponds, trout farms and fish culture units with recycling warmed water weekly. Other stocks (esp. in usual pond culture) are investigated monthly.

Health condition of fish is always to be controlled before fishing out, transporting fish and stocking. Preventive treatment can be suggested on the base of investigational results. This treatment is performed above all by the application of medicaments into the water environment and feeding by medicated feeds.

## OTHER PREVENTIVE PRINCIPLES

The ways of prevention from individual, most important viral, bacterial, fungal and parasitic diseases are described in adequate individual chapters.

Specific, very effective way of prevention from diseases is the vaccination of fish. Vaccines against following relevant viral and bacterial diseases are recently tested with different success: CCV, IPN, SVC, VHS, IHN, furunculosis, ERM, and vibriosis. Individual vaccines are applied intraperitoneally, perorally or in the form of bath. Peroral application or bath are most suitable ways from the point of view of fish culture practice. Also vaccines against some other fish diseases including parasitoses are currently developed.

Fish are subjected to therapy in those cases when a disease is so developed that the life or performance of the fish is immediately endangered or expected to be endangered in the subsequent period. Therapeutic treatment should be regarded as emergency measure resorted to when prevention has failed.

The therapeutic treatments may be as follows:

a. application of therapeutic substances and preparations to the aquatic environment (therapeutic baths for fish and eggs)
b. administration of therapeutic substances in feed
c. administration of therapeutic substances via a probe
d. administration of therapeutic substances by means of injections

Application of therapeutic substances and preparations to the aquatic environment (therapeutic baths for fish and eggs)

Therapeutic substances are put into water to control ectoparasitic, fungal and bacterial diseases of the body surface and the gills. In some cases the therapeutic baths can also be used (after absorption of the active substances via the skin) for controlling the causative agents of internal diseases.

According to the lenght of exposure, the therapeutic baths are subdivided as follows:

- immersion baths (up to 5 minutes)
- short-term baths (5 minutes to 2 hours)
- long-term baths (2 hours to several days)

General principles of therapeutic baths for fish

To perform the therapeutic baths effectively and to avoid losses of the fish, a number of general principles must be respected, including:

a) The state of health of the fish stock must be continuously monitored so that the most effective therapeutic bath can be promptly chosen and applied: fish in an advanced phase of a disease are exhausted and weak and can be easily killed by exposure to the drug in the bath.

b) The results of examination of the fish serve as a basis for determining the type of therapeutic bath. Most of the therapeutic preparations are toxic to the fish at higher concentrations, so the instructions have to be strictly adhered to.

The substances and preparations used for the baths must be fresh, packed in original containers. The dose to be used in the bath must be accurately calculated to avoid poisoning the fish by overdosage, or to avoid a poor effect if the dose is too low. If the instructions state a range of doses between two limits, then the lower amount is given to the weakened fish and the higher one to fish in good condition.

The drugs must have been dissolved before application to the water; the application itself is performed by spraying over the water surface. With the substances and preparations used for long-term therapeutic treatment of fish in reservoirs and ponds, there should be a satisfactory difference between the lethal concentration (LC) for the causative agent of the disease and the LC for the fish: the therapeutic index* is to be at least 4 or above 4, 10 at the maximum. These therapeutic means must be readily soluble in water and must easily break down.

c) Fresh and uncontaminated water must be used to prepare the solution for the bath. The physico-chemical characteristics of the water influence the effectiveness of the therapeutic substances and preparations and also their toxicity to the fish. The most important water characteristics include temperature, pH, concentration of organic substances, acid capacity (alkalinity), Σ Ca + Mg and others.

d) A tolerance test must have been conducted before any bath. The tolerance test is a bioassay on several fish to see the safety or harmfulness of the therapeutic bath for the fish stock to be treated under the existing conditions.

e) The therapeutic baths themselves are carried out in all-glass tanks, fibre-glass tubs, vats, fibre-glass plastic troughs, in concrete or earth storage basins or straight in the ponds. It is also possible to subject the fish to short-term therapeutic baths in the transport boxes during shipment if the shipment time is the same as, or shorter than, the recommended exposure time.

The fish should have been given no feed before an immersion bath or a short-term bath to avoid increased need for oxygen (for example, one to three feedings are skipped on the trout farms). Fish exposed to long-term baths, with several days' exposure times, have to be fed with supplementary feeds. Emergency scenarios must be prepared for the prevention of possible accidents: water aeration facilities must be ready for use, or precautions should be made for promptly removing the fish from the bath and putting them in fresh (preferably flowing) water, or an emergency inlet of clean and safe water must be available for fast dilution of the bath solution.

The tanks or reservoirs with the therapeutic solutions should never be overstocked: the fish must have enough space to move freely and the solution must get to every spot on the body surface of each fish. A 100-litre bath will accommodate 30 kg of fish at the maximum and the bath solution is as a rule replaced after treating 5-10 sets of fish.

* The therapeutic index says how many times the given substance's LC for fish is higher than that for the causative agent of the disease.

For long-term baths straight in the pond, the substance or preparation is either applied in a single batch into the inlet or may be evenly distributed over the water surface in the pond. For the whole period of treatment the flow of water through the pond must be stopped and warning plates should be placed around it. Residues of the therapeutic substance must have completely disappeared before water is allowed to flow through the pond again. Treatment of the whole pond is seldom resorted to: it is carried out when the fish are in acute danger. It is a problem with such large scale baths that together with the causative agents of the disease the drug used in the bath also kills the organisms in the food chain, thus reducing the nourishing capacity of the pond.

f) When the treatment is finished the fish should be removed from the bath and put into clean (preferably flowing) water. If the treatment was performed in a whole pond, the inlet source must be strong enough to allow for rapid dilution of the bath solution. All regulations and standards regarding surface water quality conservation must be respected in discharging the used therapeutic solution outside the fish culture facility. In the majority of cases the used solutions are disposed of outside the aquatic environment: for example, they are left to seep into the ground in places free of the danger of penetration into surface or underground waters.

g) The effectiveness of the therapeutic baths must be checked by macro- and microscopic examination of 5 fish at the minimum from each pond or tank after the rinsing of the treated fish in clean water. This must be done immediately after the bath, within one day of the termination of the bath at the latest.

h) It is a general principle that market fish should not be treated by therapeutic baths 14 days before shipment to the market. Treatment of market fish in malachite green bath must be avoided for 6 months before assumed time of consumption.

i) All labour safety precautions must be taken during the treatment of fish by therapeutic baths.

A survey of the most important chemicals and preparations used in the therapeutic baths of fish. Preparing and performing the baths

Sodium chloride (NaCl) is widely used in fish culture for parasite control during the rearing of the fish from the earliest stages of the fry up to the market fish. As the difference between the lethal concentrations of sodium chloride to fish and parasites is not very large, it is necessary during the treatment to stick to the general principles, especially the instructions concerning the tolerance tests. Zinc-coated containers should never be used for the NaCl baths.

Sodium chloride is largely used in the form of short-term baths which are fairly effective in the control of the species of the genera Cryptobia, Ichthyobodo, Chilodonella, Trichodina and Trichodinella, and somewhat less effective in the control of the species of the genera Dactylogyrus, Gyrodactylus, Piscicola, Argulus, and in the cases of the fungal diseases.

The salt bath is prepared by dissolving 10 to 30 g NaCl in one litre of water. The exposure time is 15 to 30 minutes. If necessary, the salt bath may be repeated in the majority of species. In the early stages of the fry, treated at a water temperature of 20 to 25°C, good results are obtained at a concentration of 10 g per litre and at an exposure time of 30 minutes.

In cyprinid culture, NaCl concentration of 20 g per litre is used at an exposure time of 15 minutes for the treatment of weaker fry. Baths of the same characteristics may also be used for the treatment of salmonids. Stronger fry and older fish of the cyprinid group may be treated with success by a bath at a concentration of 30 g per litre for 25 to 30 minutes.

It should be taken into account that at water temperatures below 5°C the effectiveness of the salt baths is substantially reduced. Sodium chloride may also be used for long-term baths (concentration of 1-2 g per litre, exposure time 1-2 days) in cases of occurrence of chilodonellosis in fish kept in storage ponds or provisional handling ponds in autumn or spring.

Formaldehyde is distributed in the form of 36-38% aquatic solution. The chemical to be used for parasite control in the fish must be a clear solution free of paraformaldehyde sediment (white sediment on the bottom). During the bath itself, the main factor to be taken into account (among the factors underlying the effectiveness and toxicity of the bath) is water temperature.

The market fish may be treated with formaldehyde bath 14 days before delivery to the market at the latest. Formaldehyde is largely used for the short-term baths to control pests of the genera Cryptobia, Ichthyobodo, Chilodonella, Trichodina, Trichodinella, Dactylogyrus, Gyrodactylus, and the fungal diseases.

The concentration of formaldehyde in the bath depends on water temperature. At water temperatures up to 10°C the concentration is 0.25 ml of 36-38 % aqueous solution per litre, at 10-15°C it is 0.20 ml per litre, and at a temperature above 15°C it is 0.17 ml per litre. The time of exposure is 30 to 60 minutes. For example, a concentration of 0.25 ml per litre and exposure time of 30 minutes at a water temperature of 25°C are recommended for the treatment of the early fry stages of cyprinids and catfish.

Long-exposure formaldehyde baths can be used in the same cases of long-exposure NaCl baths, the concentration of formaldehyde (36-38% aqueous solution) being 0.025-0.030 ml per litre. The solution is unrepeatedly applied to the water inlet and there is no time limit of exposure.

Malachite green is deep green in colour, readily soluble in water. Exposure to a therapeutic malachite green bath without prior tolerance test may kill a whole stock. Hence, every new batch of the chemical must be tested for toxicity to fish and effectiveness of parasite control before it is used for the treatment. Some limits apply to the use of malachite green in fish culture: for example, it should not be used for the treatment of the fish later than 6 month to delivery to the market (the hygiene aspect) and at the recommended concentrations and exposure times it should not be used for the treatment of the early stages of the fry (the fish safety aspect).

Malachite green is used either for the short-exposure baths or, more frequently, for long-exposure baths, especially for the control of Ichthyophthirius multifiliis and also in the cases of occurrence of the species of the genera Cryptobia, Ichthyobodo, Trichodina, Trichodinella, Chilodonella and for treatment of fish against the fungal diseases. To control the fungal diseases, it is also possible to use immersion baths in malachite green (66.7 mg per litre, exposure for 10 to 30 seconds). Recently very good results have been recorded with the use of a combined malachite green and formaldehyde bath (exposure for 2 or 6 hours).

The short-term malachite green bath uses a concentration of 6.7 mg per litre and an exposure time of 1 to 1.5 hours. At water temperatures of up to 10°C it can be performed in different types of reservoirs but when the temperature is higher it can only be done in ponds or tanks that can be drained and filled again in 30 minutes.

For the long-term malachite green bath of cyprinids, the chemical is applied at a concentration of 0.5 mg per ml, after accurate calculation of the amount of water in the reservoir or tank. For salmonids the concentration is 0.15 to 0.20 mg per 1. Upon the application of malachite green and its thorough distribution throughout the tank, the water flow is stopped and aeration is provided. Twenty-four hours later the bath is replaced: the tank is drained, clean water is left to flow through it for an hour, the tank is filled again to the same level as before and another dose of the chemical is applied. All this is done six times.

For the combined malachite green and formaldehyde bath, the water in the tank or pond should contain 0.25 mg malachite green and 0.125 ml of 36-38 % aqueous solution of formaldehyde per litre. In fibre-glass troughs the exposure time is 2 hours (with aeration provided) and in the storage ponds 6 hours.

In practice this is done as follows: the pond is drained to contain half as much water as normally, the flow is adjusted to a rate at which the water is replaced in 6 hours, and the calculated amount of solution is slowly added to the water inlet. Six hours later the flow is increased to speed up the diluting process and to increase the amount of water in the tank or pond to the normal level. When fish are treated for ichthyophphthiriasis in laminated plastic troughs, it is recommended to repeat the combined bath twice or three times in one week.

Substances and preparations containing copper are used for the therapeutic baths of fish, though they have a toxic action in the aquatic environment. $CuSO_4.5H_2O$ is used most frequently for the control of some fungal, parasitic and bacterial diseases of fish. At the present time its use is limited to the control of flexibacteriosis of the gills in the salmonids. It is used in the form of immersion bath (concentration 0.5 g per litre, exposure for 1 min) and good results are also obtained when the chemical is applied to a flow-through tank.

Another substance used for the parasite-control treatment of fish is copper in the form of oxychloride [$3Cu(OH)_2.CuCl_2.H_2O$]. It is used for short-term baths when the fish are found to harbour species of the genera Cryptobia, Trichodina, Trichodinella and Chilodonella.

This substance is also a good molluscocide, used to control aquatic molluscs, especially those of the genus Lymnea, which are intermediate hosts of the causative agent of serious fish parasitoses. Preparations based on copper oxychloride include Kuprikol 50, which contains 47.5 % of the active ingredient at the minimum. It is used at a concentration of 30-70 mg per litre for 15-30 min for the treatment of common carp and grass carp. With other fishes the therapeutic dose of Kuprikol 50 is at the level of lethal concentrations. To kill the water molluscs, Kuprikol 50 is used at a rate of 15-30 kg per ha (if the average depth of the pond is 1 m).

The therapeutic efficiency of $CuSO_4.5H_2O$ and copper oxychloride, as well as their toxicity to fish, is significantly influenced by the physical and chemical properties of the water.

Trichlorphon is used for long-term baths for cyprinids. For salmonids it is very poisonous, so it cannot be used for therapeutic baths in these species. The preparations on trichlorphon basis, used for the baths, are distributed under the brand names Masoten, Neguvon, Dipterex, Soldep and others.

When used for fish parasite control, the trichlorphon-based preparations are applied to ponds or other fish culture facilities (tanks, troughs) at a single dose. The minimum exposure is 48 hours. The parasites on the invaded fish are immobilized during the first day of treatment.

In 24 hours the intensity of invasion is considerably reduced and the percentage of immobilized parasites highly increases, in 48 hours the treatment results in a negative parasitological finding. Preparations on the basis of trichlorphon can only be applied to ponds and other facilities with perfectly tight outlet systems. During the treatment and as long as the residues of trichlorphon and its metabolite dichlorvos remain in the water, the flow through the pond must be stopped and there must be a warning plate on the pond dam.

The water flow through the pond may be resumed 2 to 3 days after getting a negative result of the bioassay: the time of persistence of the action of trichlorphon and its metabolites is determined by bioassay on daphnias. The average persistence time of these harmful substances in the pond is 1-2 weeks at a water temperature of about 20°C and water pH of 7-8, and 2-3 months in winter, when the water temperature and the pH are low. Long-continued exposure of pond water to trichlorphon and its metabolite, dichlorvos, kills the majority of the natural food for the fish.

Owing to this, full-value feeds must be administered until the natural food organisms develop again in the pond. Two weeks must have elapsed from the day of getting a negative result of the test on daphnias, before the fish may be taken from the treated pond for human consumption. The rates of administration of Soldep, containing about 25 % trichlorphon, can be used as an example.

Soldep at a concentration of 1-2.10-3 ml.litre-1, i.e. 10-20 litres per ha at an average pond depth of 1 m, is used to kill the species of the genera Dactylogyrus, Gyrodactylus, Piscicola and Argulus, and is also partly effective in the control of the genus Ergasilus. At the same concentration, Soldep also kills the intermediate hosts (Cyclops, Mesocyclops) of some fish parasites, e.g. the tapeworm Bothriocephalus acheilognathi.

The rates of administration of other organo-phosphorus preparations is proportional to the content of the active ingredient, trichlorphon. The use of trichlorphon-based preparations in ponds must always be well-thought and should only be resorted to when the fish stock is exposed to immediate danger.

Ammonia is used in combination with trypaflavin (acriflavin) in the form of immersion baths to kill the pests of the genera Dactylogyrus, Gyrodactylus and Diplozoon; the same bath may also be used when species of the genera Trichodina, Trichodinella and Chilodonella are found in the fish. The ammonia and trypaflavin baths are prepared from a store solution, which consists of 100 parts of 10 % $NH_4OH$ and one part of 2.5 % aqueous solution of trypaflavin. The solution for the bath itself is prepared by diluting the store solution with water at a rate of 1:1000.

The time for which the fish are left in the bath depends on water temperature. At temperatures up to 12°C the exposure time is 2.5 min, at temperatures above 12°C (up to 20°C) the fish are treated for only 1.5 min. No baths are performed at temperatures above 20°C. Owing to the toxicity of ammonia to fish at higher water temperatures and water pH, the use of ammonia and trypaflavin baths has been much less frequent in recent years.

Acriflavin (trypaflavin) is a brown-red crystalline powder soluble in water. The recommended therapeutic acriflavin concentrations are several times lower than the lethal concentrations to fish (the therapeutic index is about 5). For this reason, acriflavin baths can be regarded as comparatively safe to fish. Acriflavin is used in the form of long-term baths (concentration of 10 mg per litre, exposure for 10 hours), most commonly in aquarium fish culture: owing to the long exposure time, these baths are not very common in fish farming.

Acriflavin controls protozoan parasites of fish and bacterial diseases on the surface of the fish body. German authors recommend to use long-term acriflavin baths at a concentration of 3 mg per litre (exposure time 12 hours-repated three times) for the control of local flexibacterioses in trout, eel and carp.

Lime milk is prepared by dissolving 2 g of newly burnt lime in one litre of water. It is used in the form of immersion baths to kill Piscicola geometra. Carp fry are exposed to this bath for 5 seconds. For cachectic stock fish after poor hibernation the exposure time is 10 seconds and for stock carp in good condition, and for older carp, the exposure time ranges from 15 to 20 seconds. Lime bath is not recommended for fish with sensitive gills (pike, trout).

Lysol is a disinfectant aqueous solution of cresol with potassium soap. It is used at a concentration of 2 ml per litre in the form of immersion baths (5-15 seconds) to control the species of the genera Argulus and Piscicola. Lysol is not recommended for brood fish of salmonids.

Potassium permanganate is used in the form of immersion baths (1 g per litre, 30-45 seconds), short-term baths (0.1g per litre, 5-10 min; 0.01 g per litre, 60-90 min) as well as long-term baths for the control of fungal diseases, parasites (when protozooses occur) and bacterial diseases. A potassium permanganate bath at a concentration of 1 g per litre for 150 second was tested with good results for the control of Eudiplozoon nipponicum in higher age categories of carp.

This bath cannot be used at temperatures higher than 10°C. Long-term treatments are performed in storage ponds or other ponds for easy fish handling; the concentration is 0.3 to 0.6 mg $KMnO4$ per litre of water, exposure time 12 hours. The therapeutic doses of potassium permanganate are very close to the lethal concentrations to fish, so the treatment must be performed very carefully, especially with the aquarium fishes.

It should be borne in mind that in summer when the water is warm these baths may be dangerous to fish. When brood fish are handled, local injuries on their bodies are treated with a pledget or sponge soaked with potassium permanganate.

Antibiotics are recommended to be used in the form of therapeutic baths to control bacterial diseases of the skin and gills of fish. These baths are used mainly in aquaristics and today also in rearing young stages of fish in special fish culture facilities. Before the treatment, the antibiotic must be well determined as to its performance in the control of the bacteria responsible for the disease the fish suffer from. The therapeutic doses of antibiotics are in the order of tens of mg per litre at long-term baths and in the order of hundreds of mg at short baths.

Entizol, whose active ingredient is metronidazol, can be used for baths at a concentration of 4 mg per litre for 2-3 days. Metronidazol is absorbed via the gills and produces in the blood a therapeutically effective concentration to kill parasitic Flagellata, e.g. the genus Hexamita. The bath is particularly suitable for the treatment of aquarium fishes.

The method of treatment using a temporary increase in water temperature is performed by successively increasing the temperature of the water with invaded fish to 31-32°C for 3 days and then reducing the temperature again to the starting level. The fish stock gets rid of the infection and acquires an appreciable level of immunity. In fish culture practice this method is used to control ichthyophthiriasis mainly in aquarium fishes and in special warm-water fish facilities. In the rearing of the early stages of cyprinids and catfish, warming is the only efficient and practically applicable method of ichthyophthiriasis control.

## THERAPEUTIC BATHS OF THE EGGS

Malachite green, formaldehyde and sodium chloride are most frequently used in fish culture practice for the control of the fungal and bacterial diseases of fish eggs. Malachite green bath provides a good treatment of the eggs of carp, tench, sheatfish, pike, whitefish and salmonids; its concentrations range between about 5 and 10 mg per litre and exposure times are 5 to 30 minutes once to twice daily.

Malachite green is not used for the treatment of the eggs of herbivorous fishes: formaldehyde is better for this purpose, its concentration being 0.05 to 0.35 ml per litre and exposure time 10 minutes once in two hours. Formaldehyde bath can also be used for the treatment of other fishes eggs. Salmonid and whitefish eggs may also be subjected to an immersion bath of sodium chloride at a concentration of 20-50 g per litre. Acriflavin (500 mg per litre, 20-30 minutes) is also recommended for these fishes.

Besides these traditional preparations, combined-action iodine-detergent, Jodonal preparations such as e.g. Wescodyne or Incodyne have recently been used on an increasing scale: these preparations control fungi and bacteria as well as the virus diseases of fish eggs.

### Administration of Therapeutic Substances in Feed

Administration of drugs contained in feed is now practiced increasingly frequently in all types of fish culture. This approach is advantageous, hence promising, mainly from the point of view of fish farm operation. With cyprinids, the stock must have been attracted and concentrated, as far as possible, around the feeding places, and habituated to the administered feed, before the treatment itself can be started.

Administration of the same feed as normally, but containing the drugs, may be performed when there is plenty of oxygen in the water and the fish take the feed greedily. In larger water reservoirs it is difficult to habituate the fish to regular feeding, especially in those reservoirs where a larger amount of natural food is available. With salmonids it is very easy to administer drugs with feeds. Before the treatment it is recommended to skip one feeding to be sure the fish will take the medicated feed as soon and as greedily as possible.

The disandvantage is that the diseased fish take successively decreasing amounts of the feed offered to them. Heavily infected or invaded individuals do not take food at all, so the treatment has no effect on them.

The therapeutic drugs are administered either as medicated granulated feeds or are admixed to the feeds straight on the fish farm.

In the medicated granulated feeds the drug is incorporated in the pellets. The pellets are hard but they soften and swell in contact with water, where they remain compact for 12 hours. Four medicated feeds are available in Czechoslovakia at present.

These are:

VR (formerly called Karpex; the 5 kg packages distributed through pharmacies are called Rupin).

Composition: chloramphenicol palmitate 2.173 g, vitamin A 50 000 i.u., vitamin D3 25 000 i.u., methylene blue 0.3 g, saccharin 0.05 g, anise oil 0.4 g, stabilizers and obduction substances 59.68 g, wheat flour added to make 1 kg.

Indication: treatment of carp for erythrodermatitis, possibly also to control other bacterial diseases in cyprinids.

Administration: administered in the feeding place at a rate of 15 g per 1 kg of the weight of the stock per one feeding. The treatment is repeated 4 to 8 times in an interval of 2 to 3 days, depending on water temperature (two-day intervals are used when the temperature is above 20°C). The feeding must always be adjusted so as to let the fish consume the preparation within 12 hours of administration.

VR-NeO (small packages are labelled Rupin-NeO).

Composition: oxytetracyclin 1.33 g, neomycin 0.67 g, methylene blue 0.3 g, feed flour to 1 kg.

Indication: infectious diseases of cyprinids caused by oxytetracyclin- and neomycin-sensitive germs.

Administration: administered in the same way as VR.

## Taenifugin carp

Composition: niclosamide piperazine salt 7 g, obduction, auxiliary and appetizing substances 46.25 g, ground limestone 100 g, wheat flour to 1 kg.

Indication: bothriocephalosis, caviosis and caryophyleosis of cyprinids, proteocephalosis of rainbow trout.

Administration: Taenifugin carp is used in any season of the year when the fish take food. It is administered in the usual feeding places either once or repeatedly in 48 hours to allow a maximum number of the fish to take the pellets. The amount given to the fish should be equal to 1-2 % of the weight the fish stock has at the time of treatment. The efficiency of the treatment must be checked by a parasitological examination of the guts 2-4 days after the treatment. If live tapeworms are found to occur again, the treatment should be repeated.

## Chronicin salm

Composition: chloramphenicol 30.0 g, potassium propionate 100 g, feed flour to 1 kg.

Indication: furunculosis of salmonids.

Administration: Offer the medicated feed in the place where the common feed is normally given, do this daily for 7 days. The amount administered should correspond to 1 % of the weight of the stock at the time of administration. The administered dose of about 30 mg per 1 kg of fish weight is to provide the fish with an effective chloramphenicol level for 24 hours.

Other drugs administered in feed include, in particular, various antibiotics, sulphonamides, furazolidon, carboneum tetrachloratum, Entizol and others. These are admixed into the feeds just before administration. The best cyprinid feeds to carry the drugs are wheat groats or wheat flour high in gluten; for salmonids the best feed for such purposes is ground spleen or ground beef.

Antibiotics administered in feeds are used for the control of the bacterial diseases of fishes. Chloramfenicol has been the most widely used antibiotic: it has a wide spectrum of action but at the same time a number of adverse side effects, as demonstrated in recent studies. Chloramfenicol is administered at a rate of 40-60 mg per kg of live weight of the fish for 10-14 days. Its use is now declining, owing to the mentioned side effects. The best antibiotic for the treatment of each particular disease should preferably be selected on the basis of the results of antibiotic sensitivity tests in the pathogenic bacteria.

Sulphonamides can also be used with success for the treatment of bacterial diseases of fish, especially furunculosis in salmonids. They are administered in feed at a rate of 0.1-0.25 g per 1 kg of live weight of the fish for 8 days.

Furazolidon has been tested with success in the control of the bacterial diseases (especially furunculosis of salmonids and erythrodermatitis of carp) and parasitic diseases of fishes (particularly hexamitosis of salmonids and partly also coccidiosis in carp). To treat the fish suffering from bacterial diseases, furazolidon is added to the feed at a rate of 0.1-0.2 g per 1 kg of the live weight of the fish and is administered for 8 days. For the control of hexamitosis of trout the rate is 0.5 g per 1 kg of the weight of the feed and the administration is continued for 10-14 days.

Carboneum tetrachloratum ($CCl_4$) is used for the control of the most widespread spiny headed worm, Neoechonorhynchus rutili. Equal parts of $CCl_4$ and paraffin oil are added to dry feed (groats) and the rate of administration (in single treatment) is 0.5 ml $CCl_4$ per 1 kg of live weight.

Entizol (active ingredient metrinidazol) is a good therapeutic preparation to treat hexamitoses at a rate of 0.25 g per 1 kg of feed, the treatment being continued for 3 days.

Administration of therapeutic substances via a probe

This method of drug administration is resorted to in exceptional cases to treat limited numbers of fish, e.g. for the control of bothriocephalosis and caviosis in the brood fish at sites with the occurrence of these diseases, before the brood fish are transported to another area or country.

The therapeutic substance, e.g. nitrosamine piperazine salt, is dissolved in semiliquid starch gel, which is prepared by boiling about 60 g of food starch

(Solamyl) in 1 litre of water. In cyprinids the drug is administered via a thick-walled elastic hose of plastic material, connected with a syringe. The hose is introduced along the central longitudinal axis of the upper palate. The moment when the hose hits the pharyngeal teeth can be clearly identified (by feeling the mild stroke).

At this moment the hose should be inserted, with slight twisting, between the pharyngeal teeth and the crushing plate. The hose should be pushed in very lightly and only to a depth where it opens into the gullet, and then follows the administration of the drug.

## Administering therapeutic substances by injection

In the past the injection method of administration of therapeutic substances was used on mass mainly in the treatment of stock carp. Intraperitoneal administration (into the body cavity) was used mainly with chloramphenicol and later also with the vaccine against spring viraemia (prevention of the disease). However, mass use of these treatments is now becoming less common because of the great laboriousness and of the frequent mechanical injuries and stresses. The therapeutic and preventive substances are administered in feed, as far as possible.

Nevertheless, injection treatment will continue to be practiced in small groups of fish, especially the brood fish. Brood fish may receive in this way, for example, different antibiotics, vaccines, sexual hormones (in the prespawning period) and other substances; T-globulin injections are used in Poland to increase non-specific resistance of brood fish and their progenies. The drug or sexual hormone is injected into the body cavity (intraperitoneal administration) or into the muscle (intramuscular administration).

For the intraperitoneal injection, the site where the needle is to be injected is on the left side of the fish body at the point of intersection of two fictitious lines, the first starting at the base of the pectoral fin and runing along the longitudinal axis of the body and the other starting at about the centre of the pelvic fin and running perpendicularly to the first one.

The angle at which the needle is introduced into the body is also important. In the scaleless fish it should be 20 to 30 degrees, in scaly fish it should be 10 to 15 degrees, the needle passing between two successive scales. The drug flows easily from the needle introduced in the body wall, visible blotches occur under the skin.

For the intramuscular administration to the carp, the site of injection is on the left flank 1 to 2 cm behind the fore end of the dorsal fin and 3 to 4 cm below it. With other fishes the injection site is on the boundary between the first and second third of the body, 2 to 3 cm below the upper line. The needle and the injection site should be wiped with a pledget or sponge, dipped in 1 % solution of potassium permanganate.

Prevention and therapy are stated for the following major viral diseases of freshwater fishes:

## CHANNEL CATFISH VIRUS (CCV)

Prevention efforts are focused on avoiding infection of the rearing facility through strict veterinary controls on the transfers of brood material. The eggs must come from infection-free environments. Eggs at the eye point stage are preventively subjected to Jodonal bath (concentration of 2.8 - 6.7 ml per 1 litre, exposure for 5 minutes) or to a Wescodyn R bath (concentration 50 mg per 1 litre, exposure for 10 minutes).

The channel catfish fry are closely checked for health mainly during the early stage of their life until they reach a weight of 10 g. When an infection breaks out, the infected stock must be isolated, subjected to laboratory examination and - in the event of mass death - safely disposed of. Keeping the fish at a water temperature below 19°C for 24 hours may result in substantial reduction of losses. Upon mechanical cleaning, the infected environment must be disinfected with chlorinated lime (concentration of 200 - 400 ml per litre, exposure for 12 hours). All stress situations are to be avoided. Fish vaccination provides specific prevention of CCV.

No therapy has been developed as yet. Secondary bacterial infections can be controlled by trypaflavine, chlorampfenicol or oxytetracycline baths.

### Infectious pancreatic necrosis (IPN)

Prevention relies on strict veterinary controls imposed on all transfers of eggs and fish to uncontaminated fish culture facilities and the flowing waters that feed the rearing ponds. The young fry are best reared from the own stock of each farm and best kept in spring water. The start of intensive fish feeding is a critical period: the fish have to be carefully habituated with no abrupt switchover.

The preventative practices also include the general principles of hygiene and disinfection. The regular virological examination of individual trout farms is important for prevention from this disease. Fingerlings at the age of 2-4 weeks and 8-10 weeks is to subject to the examination. The possibilty of this virus registration is the highest in these periods. Fish vaccination provides specific prevention of IPN.

No therapy at a significant efficiency level has been developed as yet. The mortality of the affected stock can be reduced by adding polyvinylpyrrolidone iodine to the feed pellets at a dose of 1.6 - 1.9 g per 1 kg of the live weight of the fish per day for 15 days. This treatment can only be effective if started on time.

### Spring viraemia of carp (SVC)

Virus of SVC is spread in all fish culture facilities in our country. Prevention is to be tended to enhance the non-specific resistance of fish. Fish must be reared by well-tested methods, stress should be excluded, fish stock should be optimalized, plenty of natural food and quality replacer pellets should be provided and the fish being reared must be kept under strict health control.

It should be combined with preventative antibiotic and chemotherapeutic treatment of the fish and with the control of the causative agent through the disinfection of the fishing gears, transport containers and small reservoirs (storage ponds, wintering tanks) and waterlogged places. Ponds, reservoirs and tanks are disinfected by means of burnt lime applied at a rate of 2.0 to 2.5 tonnes per ha. Apparently diseased fish should be eliminated and suspected stocks should be reared in small ponds away from the rest of the fish farm's stock.

Fish vaccination provides specific prevention of SVC. A vaccine has been developed in Czechoslovakia; it was administered intraperitoneally on an experimental scale. From the point of view of fish farming practice, the most promising method would be the administration of the vaccine in food; this method is being currently finished in Czechoslovakia.

There is no therapy to kill Rhabdovirus carpio. Antibiotics (chloramphenicol, neomycine) or chemotherapeutics (sulphonamides), all administered in feed, are used for the prevention of secondary bacterial infections. Medicated VR or VR-Neo feeds can also be used with advantage.

### Viral haemorrhagic septicaemia (VHS)

Prevention relies on preventing the introduction of the causative agents into the fish culture facilities or to the watershed; this is done through strict veterinary controls on any transfers of eggs and stock fish. As to the general preventive measures, emphasis is laid (mainly in intensive culture facilities) on feed quality, copious supply of vitamins, hygiene of the environment and regular veterinary checkups.

It is recommended to treat the eggs by Thiomersal, Merthiolat and Bactosept baths (concentration of 0.2 mg) per litre, exposure for 10 minutes). Burnt lime (1 kg per m2), calcium cyanamide (1 kg per m2), chlorinated lime (2 kg per m2) or sodium hydroxide (2 g per litre) can be used for the disinfection of the rearing facilities and the appliances used. Fish vaccination is a specific prevention of VHS.

There is no known effective therapeutic means to control the disease. In is recommended to give the fish easily digestible feed (spleen) enriched with vitamins A, B1, B2, D and E during the critical period. Good response is also gained when the fish are given the Meso-Inosit vitamin mix at a rate of 350 mg per 1 kg of feed pellets for several days at a reduced water temperature.

## INFECTIOUS HAEMATOPOETIC NECROSIS (IHN)

It is possible to provide prevention of the disease by applying the common measures normally taken in cases of any infection. The only effective measure is to keep the stocks away from the causative agent of the disease. It is recommended to use the following practice to prevent the disease: rear the fry at a temperature of 15°C at the minimum for 1 month after hatching; then transfer the fry to normal conditions with a temperature of about 10°C. As to the eggs, it is useful to subject them to regular Jodonal baths (concentration of 4.27 ml per litre, exposure time 10 minutes) or to Wescodyn R bath (concentration of 76 mg per ml, exposure for 10 minutes). Fish vaccination provides specific IHN prevention.

No therapeutic method with a specific IHN control activity is currently available.

### Pike fry rhabdovirus (PFR)

Prevention is based on adherence to hygienic principles and implementation of infection control measures; this includes, in particular, the care of the cleanness of all gears and tools and fish culture facilities; it is also necessary to prevent water from dubious sources from entering into the fish culture environment. For prevention it is recommended to dip the eggs in the eye point stage in a Jodonal solution at a concentration of 2.85 ml per litre for 10 minutes or in Wescodyn R at a concentration of 50 mg per litre for 10 minutes.

### Swim bladder inflammation

Prevention is focused on timely identification of the foci of the infection and on adherence to the veterinary regulations (the same as for SVC). As to the general preventive measures, plenty of fresh natural food and replacer feeds should be provided mainly to carp fry. Burnt lime can be used at a dose of 2.5 tonnes per hectare for the disinfection of the fish rearing environment.

No specific-action therapy has yet been developed. Secondary bacterial infections which complicate the cases can be controlled by timely administration of antibiotics or chemotherapeutics. The therapeutic intervention can only alleviate the course of the infection and reduce the mortality of the fish. Medicated VR or VR-Neo feeds can be used with advantage, both therapeutically and preventively.

### Pox of carp

Prevention relies on the elimination of the factors underlying the occurrence of the disease. When selecting the brood fish the individuals with blisters on the body are eliminated. Lime is applied if the water in the pond or in the feed source is acid. Ectoparasites, especially the fish lice, are strictly controlled.

Indirect treatment can be performed in the pond stocked with diseased fish: apply 50 kg of burnt lime per 1 ha several (three to four) times with an interval of three to four days between each two treatments. Give plenty of feed enriched with vitamins, minerals and trace elements, in case also with chloramphenicol to the diseased fish.

### Papillomatosis of eel

Prevention relies on the veterinary checking of the eels during the first and second year after stocking (this can only be done in fishponds); if positive cases are reported, the import of elvers from the infected areas should be reduced. The import of elvers from localities free of this disease is the basic preventative pressuposition.

In practice, as a rule, no preventative measures are taken. The fish showing symptoms of the disease are eliminated from breeding and the infected individuals are killed and burnt. A description has been published concerning the treatment of fish infected with this disease, using quinine sulphate at a concentration of 60 mg per litre for several hours (quinine sulphate bath).

## BACTERIAL DISEASES

Prevention and therapy are stated for the following major bacterial diseases of freshwater fishes:

### Furunculosis

Prevention in flowing waters relies on water purity and sufficient oxygen concentration. Organic pollution of the trout and grayling waters must be as low as possible. Watersheds exposed to the danger of occurrence of furunculosis should be stocked with species less susceptible to the infection (rainbow trout) or with salmonids at a younger age (fry), which are slightly more resistant. Under such circumstances, the number of fry (fish) for the stocking is usually reduced, sometimes even to half the normal level.

If the disease breaks out, the dead fish are taken out of the water and safely disposed of (burnt or burried together with a powerful dissinfectant-chlorinated lime or burnt lime). In salmonid culture, the prevention relies on protecting the stock against the introduction of the disease with fish transferred from infected flowing waters or from fish farms where the disease occurs.

Immediately after fertilization, or at the eye-point stage, the eggs should be treated with an acriflavine solution at a concentration of 0.5 g per litre for 20 minutes. The preventative measures recommended for flowing waters must be implemented even more strictly in salmonid rearing operations where the intensity of the culture is much higher. Fish vaccination provides specific prevention of furunculosis.

Drugs to control the disease are administered either for prevention or for therapy; in both cases they are added to the food (ground spleen - admixed before administration; pellets - medication during the manufacturing process).

These drugs may include powerful antibiotics, sulphonamides, furazolidone; the medicated feed Chronicin salm is also suitable. Burnt lime (dose of 2.5 to 3.0 tonnes per ha) is used for the disinfection of the rearing facilities (small ponds). Fishing gears and tools are disinfected with a potassium permanganate (KMNO4) solution at a concentration of 10 g per litre.

### Carp erythrodermatitis (CE)

As to the available preventive measures, it is necessary first of all to handle the fish very carefully during fishing to avoid any skin injury. Try also to avoid injury during the transportation of the fish to longer distances and during stocking. Maintain the fish in good condition and in a good state of nourishment above all by support of natural food sources development in ponds. Antibiotics and chemotherapeutics may also be used for prevention.

The causative agent is very sensitive to antibiotics and sulfonamids so mass administration is the preferred route of introducing these substances in the fish body; parenteral treatment and baths are less frequent. Chloramphenicol is the most frequently used antibiotic: it is incorporated in medicated feeds, e.g. the VR brand. Oxytetracycline and neomycine may also be used: these are incorporated in the VR-NeO medicated feed. Burnt lime (2.5 tonnes per ha) or chlorinated lime (500 kg per ha) are used for the disinfection of the fish culture facilities (ponds). A 10 % solution of the lime milk is used for the disinfection of the gears and tools.

## Enteric redmouth disease (ERM)

Strict veterinary controls of transferred fish, and strict quarantine, are the most effective preventive measures. Infected fish farms are allowed to sell no other brood material but eggs, which must have been thoroughly treated with iodine preparations (e.g. Wescodyne R bath) at a concentration of 50 mg per litre for 5 - 10 minutes. Literary sources say that this is sufficient exposure to reliably kill the causative agent of the disease. Vaccination provides a specific prevention of enteric redmouth disease: the immobilized vaccine is applied in the form of bath.

ERM therapy is based on the administration of oxytetracycline or on a combination of sulphonamides with chloramphenicol. Thimethoprin combined with sulphadiazine (e.g. the Duon speciality) is considered as the most effective of the recently developed preparations. Fish therapists in some countries outside Czechoslovakia use oxolinic acid and flumequin, which are even more powerful. All these preparations are administered in feed. The problem with the treatment is that it usually has to be repeated.

## Vibriosis

Prevention of this disease relies on direct veterinary inspection of transported fish, on strictly implementing hygienic precautions in fish culture, on bacteriological analysis of the feeds prepared from sea fish, and on respecting the general conditions of the control of bacterial diseases, including quarantine, disinfection, safe disposal of diseased fish and others. Vaccination of the fish provides specific prevention of vibriosis.

The preferred therapy is the administration of sulphonamides or furazolidone in feed.

## BACTERIAL KIDNEY DISEASE

Prevention of this disease requires strict quarantine of the fish brought from sources not subjected to fish health control. If eggs are to be transported they must first be bathed in sodium methiolate at a concentration of 0.2 g per litre for 10 minutes. Jodanal can also be used (concentration of 4.3 ml per litre, exposure for 10 minutes). The eggs may be treated in this way either immediately after fertlization or later, just before transportation, when they are at the eye-point stage. Where the disease persists for a long time, sulphonamides are preventively administered in feed.

If the disease occurs, the dead fish, or the whole affected stock, must be safely disposed of (in a rendering plant). Erythromycine, administered in feed at a rate of 100 mg per 1 kg of the fish live weight for 21 days, gives the best results in the treatment of the fish suffering from this disease.

### Mycobacteriosis

Prevention is based on good hygiene of fish culture. This involves regular cleaning of the tanks and avoidance of overstocking. Plenty of varied food should be available to the fish to maintain them in good condition. Stress situations (control fishing, transportation) must be minized and long-term weakening of the fish (caused e.g. by inadequate temperatures or physico-chemical properties of water) must be avoided. Suspect individuals should be immediately removed.

When fish are purchased and especially when they are imported from a foreign country, the supplier should be required to provide a veterinary certificate proving that the fish came from a disease-free environment. Nevertheless, in spite of the certificate, the purchased fish must spend some time in quarantine.

Medicamentous treatment in advanced stages of mycobacteriosis usually fails to bring longer-lasting results. However, longer exposure to streptomycin baths in an earlier stage of the ailment can stop the disease: add 50 to 300 mg of the drug per 1 litre of water and leave the fish in the bath for 4 to 6 days. In cases of mass infection it is recommended to remove and destroy all the fish stock and the plants and disinfect the tank and everything inside it (preferably using preparations containing active chlorine).

### Myxobacteriosis

This is a group of diseases of which the following three are the most serious:

#### *Columnarosis*

Prevention of this disease involves careful handling of the fish during fishing, grading and transport. Ectoparasite control is another important factor of prevention. In salmonid culture it is useful to reduce water temperature to a level at which the development of the disease is curbed.

Therapy based on baths is most frequently used in salmonids and in aquarium fishes, using the advantage of the fish being kept in small containers. Immersion bath is used for the treatment of salmonids, the fish being dipped in a solution of blue vitriol (CuSO4.5H2O) at a concentration of 0.5 g per litre for 1 minute. Malachite green is also a good chemical for immersion baths (concentration of 66.7 mg per litre, exposure time 10 to 30 seconds). Sulphonamides are administred in feeds to the fish suffering from columnarosis. Longer-term baths are used in aquarium fish culture, e.g. chloramfenicol at a concentration of 60 mg per litre for six days.

### *Cytophagosis*

The therapy is about the same as with columnarosis, except for the recommended reduction of water temperature: for the treatment of cytophagosis it is recommended to increase water temperature. An increase in water temperature above 15°C will suffice to arrest the disease. However, this treatment is difficult to provide in salmonids in many cases.

### *Bacterial gill disease*

Prevention of this disease relies on maintaining good hygienic and breeding conditions. The food must contain plenty of vitamins of group B. Trout of the critical size have to be reared in clean water (preferably spring water). The size of the stock must be adjusted according to water purity and water inflow rate. All debris and dirt must be removed from the rearing tanks (especially the residues of food and the excrements). A small amount of cod liver oil may be added to the feed replacer if it is too dusty.

Stocks in chronically diseased rearing facilities are preventively treated in critical periods of the year, or soon after the first signs of the disease occur. The treatment rests in short term or dipping (immersion) baths; during the short term bath an aeration device must be used to maintain sufficient oxygen in the water for the fish. Regularly performed baths in substances containig active chlorine (e.g. chloramine in concentration of 20 mg.l-1 for 20 min.).

Immersion baths in blue vitriol (CuSO4.5H2O) give good therapeutic results (concentration of 0.5 g per litre, exposure time 1 to 1.5 minutes). Malachite green is another good chemical for short-term bath treatment. Also sulphonamids and chinolone chemotherapeutics respectively can be applied by peroral way.

## FUNGAL DISEASES

### Oomycetosis (primarily caused by Achlya and Saprolegnia)

The best of the therapeutic substances is malachite green B (oxalate compound), active even at a very low concentration. The amount used for the dipping treatment is 66.7 mg per one litre of water, the exposure time being 10-30 seconds. For a short bath with an exposure time of 1-1.5 hours the concentration is also 6.67 mg of MG per one litre of water. For a long-time bath (6 days), the concentration must be much smaller, 0.2 to 0.5 mg per one litre of water.

The effectiveness of the long-term bath depends on water quality. The fish usually have to undergo the bath again three days later. This approach is suitable for the treatment of fish in large storage basins or tanks. For aquarium fishes and salmonids it is recommended to use the lowest active concentration of malachite green, 0.15 mg per litre.

Market fish should be free of malachite green treatment 6 months before going to the market. The eggs of salmonids can be subjected to preventive and therapeutic treatment at a rate of 5 mg MG per one litre of water for an hour twice a week. This is done without handling the eggs, the store solution being left to drop into the water that flows to the tank.

Saprolegniasis of fish can also be controlled by short potassium permanganate baths. The concentration of the chemical in the bath is 0.01 g per 1 litre of water and exposure time is 1 to 1.5 hours.

## Branchiomycosis

Preventive treatment is conducted in ponds where branchiomycosis permanently occurs in July and April, when burnt lime powder is applied onto the water surface at a rate of 50-100 kg per hectare. It is recommended in this period to increase the rate of flow through such ponds to improve oxygen content.

It is also advisable to avoid putting too much organic matter in the pond through intensive organic manuring and intensive breeding of water fowl. Burnt lime should be used at a rate of 2.5 t per hectare to dissinfect the bottom. If the disease occurs in its acute stage, stop feeding the fish, increase the flow through the pond and treat the water surface with chlorinated lime at a rate of 10-15 kg per ha (at an average pond depth of 1 m) three times a week at the maximum. When the branchiomycosis is over it is recommended to administer medicated feed (with antibiotics) to prevent secondary bacteriosis of the gills.

## Ichthyophonosis

No therapy is available as yet. Prevention is based on thorough veterinary inspection of the fish rearing facilities and quality control of the feed on trout farms where sea fish are administered. The preventive sanitation measures include pasteurization of potenntially infected fish flesh. The reservoirs infested with the fungus are disinfected with burnt lime at a rate of 2.5 t per hectare. Aquarium tanks are drained from above and the bottom and walls are dissinfected with potassium permanganate at a rate of 1 g per 100 litres of water for 90 minutes.

## Moniliales

In exophialosis prevention, increased attention should be paid to the quality of food especially the tubificids, which are vectors of Exophiala pisciphila to aquarium fishes. The most vulnerable of the aquarium fishes is Hemmigramus pulcher. In ochroconiosis neither prevention nor effective therapy are available yet. In fusariosis prevention and therapy are similar to those in coelomycetosis.

## Coelomycetes

Prevention of this disease is based on maintaining the water at an optimum quality. In places where it occurs the fungi must have been killed in the reservoirs or tanks (by disinfection repeated twice or three times with 20-day intervals between the treatments) before stocking and during the rearing. The disinfectant suitable for the treatment of tanks without fish include a solution of sodium hypochloride with a 2 % content of active chlorine, or a 2 % formaldehyde solution, the exposure time being 20-30 minutes.

# PARASITIC DISEASES

The methods of prevention and therapy of important parasitic diseases are described as follows.

## Protozooses

1. Fish diseases caused by ectoparasitic Protozoa

## Ichthyobodosis

Prevention of ichthyobodosis should be realized with absolutely full care. The main effective rules of prevention are:

a) To use the source of water free of parasites and their cysts.

The underground water is the most suitable source of water free of parasites. Nevertheless, these sources are very limited even for trout farms, hatcheries and/or other special fish culture facilities. In most cases, water from streams is used as a source of inlet water. In such cases, suitable filters can free the inlet water from infectious stages of parasites prior the water enters smaller tanks with intensive fish culture. A water from ponds stocked with fish is absolutely unsuitable to be used in trout farms, hatcheries and/or rearing facilities with early stages of fishes.

b) The stocking of healthy fish

Only well examined fish, free of ichthyobodosis and treated by antiparasitic bath (in case of necessity) should be stocked into ponds and fish rearing facilities.

c) Creation of presuppositions for good performance and health conditions of fish.

It refers particularly to the optimum quality of water without physical and chemical stressing effects. It is necessary to avoid the stress effects caused by other factors too, namely by inconsiderate handling with fish. Through appropriate culture management, the maximum development of natural food should be gained.

The supplementary feed must be applied in sufficient quantity and quality. Abundant natural food of suitable species and size composition is a basic preventive measure in rearing the early stages of fry. The maximum level of stock density must be calculated carefully. Unreasonably dense stock in trout ponds and/or raceways, in fry reared in special culture facilities, in any other fish production units, and in ponds too, leads to stressing situations, worsens the nutritional state of fish, decreases the resistance and opens the door to infections.

d) Regular control of health condition of fish

A preventive control of health condition of fish reared in hatcheries and in special fish culture facilities (in case of early fry rearing) must be realized twice a week, in trout farms once a week. A control of health condition must be preformed just prior the fishing, the fish transportation and/or stocking the fish.

Therapy - Many antiparasitic baths and their modifications have been proposed as a therapeutic measure. The effectiveness of baths is not absolute; some flagellata survive in folds of epithelium which creates new presupposition for infection. Basicaly, short-term baths in sodium chloride and/or formaldehyde are used for ichthyobodosis control treatment in early stages of cyprinids and European catfish, reared in facilities utilizing the heated effluents.

In salmonid fish culture, a long-term bath in malachite green is recommended in most cases from practical aspects. In cyprinids and salmonids, a two-hour- and/ or six-hour-long combined bath in malachite green and formaldehyde can be used too. A long-term bath in acriflavine is recommended in aquaria fish culture for ichthyobodosis control.

## Cryptobiosis

Abundant natural food of suitable species and size composition preventing the condition impairing of fish is a basic preventive measure in rearing the early stages of fry and preventing the protozoal infection. Optimum quality of water is an another important aspect. In case of cryptobiosis, a frequent sludge removing from rearing tanks is a necessary measure preventing the increase in organic substances concentration and level of organic detritus in water.

To control the dense growths of Cryptobia on the body and the gill surface, a decrease of organic load and of organic detritus (by frequent sludge removing from rearing tanks) together with improving the quality of nutrition (by abundant natural food) is mostly sufficient. In case this measurement is not effective, the same baths as in ichthyobodosis can be applied.

## Ichthyophthiriosis

Prevention is based both on the stocking of healthy fish and the use of water free from parasite infectious stages. In special fish culture facilities with heated water, in trout culture, and in special wintering ponds, inlet water from lower layers and/or from retention reservoirs without fish stock should be used. In case of inlet water from streams, water must be filtered through gravel-sand filters. If diseased fish have appeared in fish culture facilities or ponds, tanks and pond bottom must be completely dried after fishing and disinfected with calcium oxide (2.5 t.ha-1) and/or calcium hypochlorite (0.5 - 0.6 t.ha-1).

The fishing equipment contacted with water and/or fish from infected environment, must be thoroughly dried. It is important to take into mind when controlling ichthyophthiriosis, the encysted parasites (tomonts) can survive in aquatic environment 1 week in water temperature 20°C and 2 weeks in low temperatures. After this period, aquatic environment is free from parasites.

The creation of optimum living conditions is another important presupposition ensuring the good health state of stocks in course of their culture. In case of ichthyophthiriosis prevention, a respective good nutritive and health state of fish is a presupposition together with appropriate density of stock, particularly in early fry rearing and in wintering ponds. Unreasonable dense stock enables the spreading of infection. When fish are kept in good condition, parasites survive in minimum numbers. In weakened fish, in decrease of their resistance and/or in increase of water temperature (which is common in spring period in wintering ponds), parasites reproduce intensively.

Therapy of ichthyophthiriosis is difficult. Due to immersing into the surface tissues of the fish body, parasites are more or less protected against the antiparasitic baths. A method of temporary increase of water temperature and antiparasitic bath can be used in therapeutic treatment of ichthyophthiriosis. Method of temporary increase of water temperature to 31 - 32°C for 3 days is well proved in aquaria fish culture and in special fish culture facilities utilizing the heated water.

Antiparasitic baths must be used in period when adult trophonts leave the surface of host to produce a cyst. Baths are effective to these free stages of parasite only. Most frequently, long-term bath in malachite green is applied. Based on specific culture conditions, short-term baths in malachite green can be used too. Recently, a repeated combined bath in malachite green and formaldehyde for 2 hours has been proved with success in aquaria and troughs, and for 6 hours in earth ponds and holding ponds.

## Chilodonellosis

Prevention of chilodonellosis is based in creating the optimum conditions of fish culture and in preventing the transfer of causative parasite. To prevent the infestation of rearing tanks, the diseased fish are not allowed to stock and gravel-sand filters are recommended to install on inlets to special fish culture facilities. Optimum quality of water is important as well; low content of organic substances and high content of oxygen in water is unfavourable for parasite. An appropriate nutrition of fish, particularly prior the wintering, is an important preventive measure.

Therapy of chilodonellosis is realized through several antiparasitic baths which, unfortunately, are not very effective in certain cases. Short-term baths in sodium chloride, formaldehyde, malachite green and bath in Kuprikol 50 as regards to carp and grass carp can be mentioned. As to long-term baths suitable for chilodonellosis control, bath in malachite green (fish kept in holding ponds) and long-term bath in sodium chloride and formaldehyde can be applied. A long-term bath in acriflavine is used in aquaria fish culture.

## Trichodinosis

Prevention is identical with measure for skin and gill protozooses.

Therapy is identical with chilodonellosis.

2. Fish Diseases Caused by Endoparasitic Protozoa

## Trypanoplasmosis, trypanosomosis

Prevention is based in diminishing and control of leeches occurrence. A good condition of fish prior the wintering is important to diminish the losses caused by blood flagellata.

Effective therapy has not been worked out yet.

## Hexamitosis

Prevention - To prevent the transfer of cysts into the fish culture facility via both infected fish and inlet water is one of the most important measures. Regular control of health state of fish and microscopic examination of fish excreta (once in a week in case of salmonid fry rearing) is a necessity. In case of positive findings, a preventive dose of furazolidon (0.3 g per 1 kg of feed) is applied daily for 14 days. Appropriate hygienic conditions and sufficient nutrition of fish belong to important preventive measures in hexamitosis control.

Therapy - furazolidon in feed can be used successfuly for therapeutic treatment. The Polish preparation Entizol, applied in feed, is also a very effective drug. This preparation can be used in form of long-term bath as well.

## Coccidiosis

Prevention is based on oocysts and sporocysts (occurring on the bottom of ponds and holding ponds) control. Drying the bottom and its disinfection is very effective. Regular wintering, drying and disinfection of muddy parts of nursery ponds with burnt lime (2.5 t.ha-1) and/or chlorinated lime (0.5 - 0.6 t.ha-1) is an effective method of prevention. To prevent the transfer of parasite from older classes of fish and/or from coarse fish to carp fry is an another important element of prevention.

In natural spawning of carp, brood fish should be fished out as soon as possible from spawning ponds. In infected cyprinid culture, supplementary feeds must be applied frequently to prevent the fry to feed on benthos due to abundant population of coccidia oocysts.

Therapy - peroral application of furazolidon in dose 0.3 g per 1 kg of feed for 5 days is used in carp fry rearing.

## Myxosporeosis

Prevention is focussed on regular veterinary examination of fish and on selection of fish with positive findings out of culture. Most of species of freshwater myxosporea have a very resistant spores, some of them are adapted to temporal drying-up. In natural environment, spores keep their viability for 6 months to 2 years based on different conditions. Soon or later, spores sediment on the bottom, and mud and benthos create a source of infestation. Due to turbulent stream movement, spores are transferred to other localities. Spores can be distributed by fish-consuming birds too (spores were not damaged in bird's digestive tract).

Myxosporeosis control must be focussed on spores as the infectious stage, the only able element in the developmental cycle. Experiments with myxosporeosis therapy failed yet so the spores control is the only measure. Pond bottom disinfection through summer drying and chemical substances application is very important. Optimum method of prevention includes the drainage of pond in the autumn period and application of disinfective substance on the wet bottom.

Afterwards, pond is filled with water only to cover the bottom, and disinfective substance is let to act for 3 weeks. Then the pond is flushed-through, drained and dried through the winter period. It is recommended to repeat this process once again in spring time, and then the pond can be filled with water. Calcium cyanamide in dose 0.5 kg per 1 m2 of bottom area is advised as an optimum disinfection.

Therapy of myxosporeosis is not worked out yet.

### Myxobolosis

Prevention is focussed on regular veterinary control of fish health state and on control of spores, as it is mentioned in myxosporeosis prevention. Rearing the fry till the age of 2 to 3 months in plastic troughs and/or concrete tanks with inlet of suitable water is another measure in prevention. After this period, fry can be stocked in earth ponds. In ponds permnanently watered from naturally disease-infected localities, rearing the fry till the age of 2 to 3 months is not realized.

## METAZOOSES

### Monogeneosis

Prevention is based on excluding the parasites transfer from culture environment via both diseased fish and inlet water including the infectious Dactylogyrus and Eudiplozoon larvae. Natural spawning should be restricted and artificial propagation of fish should be utilized which excludes the contact of brood fish and fry. It must be avoided to stock the infected fish into the rearing facilities.

Each fish species must be stocked as a single age-class category into ponds. In nursery ponds, inlet water should be free from infected larvae, i.e. source of this water is out of reservoirs stocked with the same fish species. Abundant natural food and supplementary feeding the fry with quality feeds is another preventive measure in carp fry rearing. In such case, 6 cm body length of fry is reached as soon as possible, and in this stage, the fry is relatively resisting to Monogenea (particularly to Dactylogyrus vastator). Till this moment, a regular frequent examination of health state of fish is necessary.

Therapy. Short-term formaldehyde bath in the course of re-fishing the fish, and long-term bath of cyprinids in trichlorphone-based preparations diluted directly in pond, is an effective therapeutic method in dactylogyrosis and gyrodactylosis. It is necessary to note that trichlorphone application in ponds is followed by intensive reduction in natural food abundance.

It is therefore necessary to feed the fish intensively with high-quality feeds till the time of re-development of natural food. In case of lower intensity of carp fry infection in nursery ponds, a standard measure to improve the culture environment and health state of fish is applied instead of trichlorphone application.

Inlet of oxygen-rich water is increased, rotten vegetation is removed, and fry is fed intensively. Short-term bath in sodium chloride can be mentioned too, nevertheless it is a less used treatment method with no extra effectivity. Ammonia and trypaflavine immersion bath is too risky in conditions of higher temperature and pH of water.

Potassium permanganate bath in concentration 1 g.1-1 for 150 sec is used as therapeutic method in eudiplozoonosis. This bath is recommended for older carp and it is effective up to water temperature 10°C.

## Cestodoses

This group of diseases is represented mainly by the following five parasitoses:

### Khawiosis

Prevention is based on careful veterinary control of transported fish and fish reared in own culture facilities. The preventive feeding with medicated feed Taenifugin-carp is applied in case of low intesity infection in fry and stock carp. This preventive feeding is obligatory in case of transportation of fish from infected localities to areas free of parasite. As to antiparasitic treatment in brood fish, piperazine salt of niclosamide is applied in starch gel through probe.

When khawiosis is registered in growing period, intensive feeding can prevent the fish to consume the tubificids with developmental stages of parasite. Direct control of intermediate host tubificids is not realised in our conditions. Abundance of tubificids is controlled through pond reclamation, winter and/or summer drying of pond bottom, and bottom disinfection (2.5 t.ha-1 of burnt lime or 0.5 - 0.6 t.ha-1 of chlorinated lime). This is also a method to control the occurrence of Khawia sinensis eggs.

Therapy. Single or repeated application of medicated feed Taenifugin-carp is proved as a therapeutic treatment.

### Caryophyllaeosis

Prevention is identical as to khawiosis.

Therapy is identical as to khawiosis. It can be applied in pond culture only.

### Bothriocephalosis

Prevention is also identical as to khawiosis. A preventive control of intermediate hosts is not realized. Only in extreme mass occurrence of bothriocephalosis, application of trichlorphone-based preparations (e.g. Soldep in dose 10 l.ha-1 for a water column 1 m on average) is used to control the intermediate hosts in a reservoir stocked with fish.

Therapy is identical as to khawiosis.

## Proteocephalosis

Prevention is applied in cage culture of rainbow trout only. It is identical with preventive measure in khawiosis and bothriocephalosis. First of all, it refers to careful control of health state of fish, prevention the spreading of parasitosis into other localities, and preventive feeding the rainbow trout with feed medicated with piperazine salt of niclosamide.

Therapy. Single or repeated feeding the trout with feed medicated with piperazine salt of niclosamide in dose 5 g per 1 kg is applied as a therapeutic measure.

## Triaenophorosis

Prevention is based on excluding the infected pike from inlet water entering the trout culture.

Therapy is not practised.

## Trematodoses

Prevention is based on regular examination of health state of fish. In more intensive infestation, interruption of developmental cycle of parasite is realized by means of control of intermediate hosts. In diplostomosis and postodiplostomosis, occurrence of fish-consuming birds as definite hosts should be restricted in intensive fish culture.

This can be done through a control of gull eggs, a use of nets over the trout raceways etc. Snails are controlled by biological, physical and/or chemical method. No any method itself is sufficiently effective. To interrupt the developmental cycle of parasite, individual methods must be combined.

In herbivorous fish fry rearing in ponds with intensive occurrence of snails, careful drying and winter freezing of pond bottom is recommended. Afterwards in spring period, a pond is partly filled with water. After leaving the bottom mud, pond snails (Lymnaea stagnalis) are controlled by means of chemicals.

The preparations with copper as an active substance (e.g. Kuprikol 50 in dose to 30 kg.ha-1 for a water column 1 m) are effective to snails. Other molluscocides (e.g. based on tin) are tested too. The pond is filled with water through a sieve (mostly made from synthetic web called Uhelon) installed directly on an inlet pipe. The sieve must be regularly controlled and cleaned. This is a very simple way preventing the snails to reach the pond.

These mechanical, physical and chemical methods of snail control can be suitably combined with biological method. In such case, 500 - 600 individuals per 1 ha of two-year-old carp (over 500 g) and/or three-year-old tench are stocked into pond. This stock frequently consumes small juvenile snails. In the USSR, black carp is stocked in water bodies to control large quantities of intermediate host snails.

Fry of herbivorous fish fed in troughs for several days prior the stocking is stocked into ponds at the turn of June and July. In this period, the stock carp and tench are fed with highly effective feed rich in animal protein to prevent the losses.

Therapy is not yet realized in practice.

### Acanthocephalosis

Prevention is based on regular examination of health state of stocked fish and even fish from flowing waters from which the ponds are fed. A main aim of prevention is to prevent the introduction of parasites. Careful drying and/or disinfection of pond bottom with 2.5 t.ha-1 of burnt lime or 0.5 - 0.6 t.ha-1 of chlorinated lime applied immediately after the fishing, controls the eggs and intermediate hosts. Recently an increase in pond fish culture intensity seems to be an important limiting factor of carp neoechynorhynchosis occurrence. It was recognized that an occurrence of intermediate hosts is very sporadic in intensively managed ponds with dense stocks.

Therapy was worked out against neoechinorhynchosis of carp. Carboneum tetrachloratum and paraffin oil in the same ratio is applied in a single dose 1 ml per 1 kg of fish in feed.

## FISH DISEASES CAUSED BY PARASITIC CRUSTACEANS

This group of diseases is represented mainly by the following parasitoses: ergasilosis, lernaeosis and argulosis.

Prevention is based on the excluding infected fish, both from culture and coarse water, to enter the fish culture facilities. Water sources used to supply the fish culture facilities must be under strict control. Thorough drying and/or disinfection of pond bottom with burnt lime or chlorinated lime is a method controlling the eggs and larvae of parasites.

Therapy of ergasilosis and lernaeosis is very difficult. Long-term bath in preparations based on trichlorphone proved to be only temporary effective on nauplia and copepodite stages under higher water temperature (20°C); the same bath is uneffective on adult parasites if applied under lower water temperature. Immersion bath in lysol is recommended for re-fished fish. Long-term bath in preparations based on trichlorphone can be applied directly in pond to cyprinids as argulosis treatment.

### Hirudineosis

Prevention - disinfection of pond bottom with burnt lime (2.5 t.ha-1) controls the occurrence of leeches (Piscicola geometra). To prevent the entering of infected coarse fish into the fish culture environment is also an effective method. Another way of prevention consists of creation the unsuitable conditions for leeches in pond which means an eradication of hard littoral vegetation, stones and tree branches serving as a shelter for leeches and/or a substrate for cocoons with eggs.

Therapy is applied primarily in cyprinids. A single application of trichlorphone-based preparation directly into pond is an effective method of leeches control. Blue vitriol ($CuSO_4.5H_2O$) in dose 0.05 mg.1-1, i.e. 0.5 kg.ha-1 of pond for water column 1 m on average, applied 3 - 4 times repeatedly in a two- to four-week period is recommended as a control of leeches in carp ponds by some authors.

An immersion lime and/or lysol bath, and as the case may be, short-term bath in sodium chloride can be used for leeches control in re-fished fish. It must be taken into mind, that leeches drop off from fish after bath and they are able to recover again in suitable conditions. The salt bath is less effective. From brood fish, the leeches can be removed mechanically by pincette.

## POPULATION DYNAMICS OF FISHERIES

A fishery is an area with an associated fish or aquatic population which is harvested for its commercial or recreational value. Fisheries can be wild or farmed. Population dynamics describes the ways in which a given population grows and shrinks over time, as controlled by birth, death, and emigration or immigration. It is the basis for understanding changing fishery patterns and issues such as habitat destruction, predation and optimal harvesting rates. The population dynamics of fisheries is used by fisheries scientists to determine sustainable yields.

The basic accounting relation for population dynamics is the BIDE model:

N1 = N0 + B ? D + I ? E

where N1 is the number of individuals at time 1, N0 is the number of individuals at time 0, B is the number of individuals born, D the number that died, I the number that immigrated, and E the number that emigrated between time 0 and time 1. While immigration and emigration can be present in wild fisheries, they are usually not measured.

A fishery population is affected by three dynamic rate functions:

- Birth rate or recruitment. Recruitment means reaching a certain size or reproductive stage. With fisheries, recruitment usually refers to the age a fish can be caught and counted in nets.
- Growth rate. This measures the growth of individuals in size and length. This is important in fisheries where the population is often measured in terms of biomass.
- Mortality. This includes harvest mortality and natural mortality. Natural mortality includes non-human predation, disease and old age.

If these rates are measured over different time intervals, the harvestable surplus of a fishery can be determined. The harvestable surplus is the number of individuals that can be harvested from the population without affecting long term stability (average population size). The harvest within the harvestable surplus is called compensatory mortality, where the harvest deaths are substituting for the deaths that would otherwise occur naturally. Harvest beyond that is additive mortality, harvest in addition to all the animals that would have died naturally. Care is needed when applying population dynamics to real world fisheries. Over-simplistic modelling of fisheries has resulted in the collapse of key stocks.

# POPULATION SIZE

The population size (usually denoted by N) is the number of individual organisms in a population. The effective population size (Ne) was defined by Sewall Wright, who wrote two landmark papers on it (Wright 1931, 1938). He defined it as "the number of breeding individuals in an idealized population that would show the same amount of dispersion of allele frequencies under random genetic drift or the same amount of inbreeding as the population under consideration". It is a basic parameter in many models in population genetics. Ne is usually less than N (the absolute population size).

Small population size results in increased genetic drift. Population bottlenecks are when population size reduces for a short period of time. Overpopulation may indicate any case in which the population of any species of animal may exceed the carrying capacity of its ecological niche.

## Virtual population analysis

Virtual population analysis (VPA) is a modelling technique commonly used in fisheries science for reconstructing historical fish numbers using information on death of individuals each year. This death is usually partitioned into catch by fisheries and natural mortality.

VPA is the most commonly used term to refer to cohort reconstruction techniques used in fisheries. It is virtual in the sense that the population size is not observed or measured directly but is inferred or back-calculated to have been a certain size in the past in order to support the observed fish catches and an assumed death rate owing to non-fishery related causes.

## Minimum viable population

The minimum viable population (MVP) is a lower bound on the population of a species, such that it can survive in the wild. More specifically MVP is the smallest possible size at which a biological population can exist without facing extinction from natural disasters or demographic, environmental, or genetic stochasticity. The term "population" refers to the population of a species in the wild.

As a reference standard, MVP is usually given with a population survival probability of somewhere between ninety and ninety-five percent and calculated for between one hundred and one thousand years into the future.

The MVP can be calculated using computer simulations known as population viability analyses (PVA), where populations are modelled and future population dynamics are projected.

## MAXIMUM SUSTAINABLE YIELD

In population ecology and economics, the maximum sustainable yield or MSY is, theoretically, the largest catch that can be taken from a fishery stock over an indefinite period. Under the assumption of logistic growth, the MSY will be exactly at half the carrying capacity of a species, as this is the stage at when population growth is highest. The maximum sustainable yield is usually higher than the optimum sustainable yield.

This logistic model of growth is produced by a population introduced to a new habitat or with very poor numbers going through a lag phase of slow growth at first. Once it reaches a foothold population it will go through a rapid growth rate that will start to level off once the species approaches carrying capacity. The idea of maximum sustained yield is to decrease population density to the point of highest growth rate possible. This changes the number of the population, but the new number can be maintained indefinitely, ideally.

MSY is extensively used for fisheries management. Unlike the logistic (Schaefer) model, MSY in most modern fisheries models occurs at around 30% of the unexploited population size. This fraction differs among populations depending on the life history of the species and the age-specific selectivity of the fishing method.

However, the approach has been widely criticized as ignoring several key factors involved in fisheries management and has led to the devastating collapse of many fisheries. As a simple calculation, it ignores the size and age of the animal being taken, its reproductive status, and it focuses solely on the species in question, ignoring the damage to the ecosystem caused by the designated level of exploitation and the issue of bycatch. Among conservation biologists it is widely regarded as dangerous and misused.

### Recruitment

Recruitment is the number of new young fish that enter a population in a given year. The size of fish populations can fluctuate by orders of magnitude over time, and five to 10-fold variations in abundance are usual. This variability applies across time spans ranging from a year to hundreds of years. Year to year fluctuations in the abundance of short lived forage fish can be nearly as great as the fluctuations that occur over decades or centuries. This suggests that fluctuations in reproductive and recruitment success are prime factors behind fluctuations in abundance. Annual fluctuations often seem random, and recruitment success often has a poor relationship to adult stock levels and fishing effort. This makes prediction difficult.

The recruitment problem is the problem of predicting the number of fish larvae in one season that will survive and become juvenile fish in the next season. It has been called "the central problem of fish population dynamics" and "the major problem in fisheries science". Fish produce huge volumes of larvae, but the volumes are very variable and mortality is high. This makes good predictions difficult.

According to Daniel Pauly, the definitive study was made in 1999 by Ransom Myers. Myers solved the problem "by assembling a large base of stock data and developing a complex mathematical model to sort it out. Out of that came the conclusion that a female in general produced three to five recruits per year for most fish."

## Shifting baseline

Shifting baseline (also known as sliding baseline) is a term used to describe the way significant changes to a system are measured against previous baselines, which themselves may represent significant changes from the original state of the system. The term was first used by the fisheries scientist Daniel Pauly in his paper "Anecdotes and the shifting baseline syndrome of fisheries". Pauly developed the term in reference to fisheries management where fisheries scientists sometimes fail to identify the correct "baseline" population size (e.g. how abundant a fish species population was before human exploitation) and thus work with a shifted baseline. He describes the way that radically depleted fisheries were evaluated by experts who used the state of the fishery at the start of their careers as the baseline, rather than the fishery in its untouched state. Areas that swarmed with a particular species hundreds of years ago, may have experienced long term decline, but it is the level of decades previously that is considered the appropriate reference point for current populations. In this way large declines in ecosystems or species over long periods of time were, and are, masked. There is a loss of perception of change that occurs when each generation redefines what is "natural".

Most modern fisheries stock assessments do not ignore historical fishing and account for it by either including the historical catch or use other techniques to reconstruct the depletion level of the population at the start of the period for which adequate data is available. Anecdotes about historical populations levels can be highly unreliable and result in severe mismanagement of the fishery.

The concept was further refined and applied to the ecology of kelp forests by Paul Dayton and others from the Scripps Institution of Oceanography. They used a slightly different version of the term in their paper, "Sliding baselines, ghosts, and reduced expectations in kelp forest communities".

The term has become widely used to describe the shift over time in the expectation of what a healthy ecosystem baseline looks like. In 2002, filmmaker and former marine biologist Randy Olson broadened the definition of shifting baselines with an op-ed in the Los Angeles Times. He explained the relevance of the term to all aspects of change, and the failure to notice change in the world today. He and coral reef ecologist Jeremy Jackson (of Scripps Institution of Oceanography) co-founded The Shifting Baselines Ocean Media Project in 2003 to help promote a wider understanding and use of the term in discussions of general conservation.

A conceptual metaphor for a shifting baseline is the price of coffee. A cup of coffee may have only cost a $0.05 in the 1950's, but in the 1980's the cost shifted to $1.00 (ignoring inflation). The current (21st century) coffee prices are based on the 1980s model, rather than the 1950s model. The point of reference moved.

The Shifting Baselines Ocean Media Project grew from its three founding partners (Scripps Institution of Oceanography, The Ocean Conservancy, and Surfrider Foundation) to over twenty conservation groups and science organizations. The project has produced dozens of short films, public service announcements, and Flash videos along with photography, video, and stand-up comedy contests, all intended to promote the term to a broader audience. The Shifting Baselines Blog, "the cure for planetary amnesia" is run by the Shifting Baselines Ocean Media Project on the Seed (magazine) Science Blogs.

## FISH STOCK

Fish stocks are subpopulations of a particular species of fish, for which intrinsic parameters (growth, recruitment, mortality and fishing mortality) are the only significant factors in determining population dynamics, while extrinsic factors (immigration and emigration) are considered to be insignificant. All species have geographic limits to their distribution. which are determined by their tolerance to environmental conditions, and their ability to compete successfully with other species. In marine environments this may be less evident than on land because there are fewer topographical boundaries, however, discontinuities still exist, produced for example by mesoscale and sub-mesoscale circulations that minimize long-distance dispersal of fish larvae .

For fishes, it is rare for an individual to reproduce randomly with all other individuals of that species within its biological range. There is a tendency to form a structured series of discrete populations which have a degree of reproductive isolation from each other in space, in time, or in both. This isolation is reflected in the development between sub-populations of genetic differences, morphological variations and exposure to different chemical regimes and parasitic species. Sub-populations also respond to fishing in such a way that fishing on one population appears to have no effect on the population dynamics of a neighbouring population.

The currently accepted definition of a stock in fisheries science, is that of Begg et al. (1999), "...[a "stock"] describes characteristics of semi-discrete groups of fish with some definable attributes which are of interest to fishery managers.". Stock identification is a field of fisheries science which aims to identify these subpopulations, based on a number of techniques.

The United Nations defines straddling stocks as "stocks of fish such as pollock, which migrate between, or occur in both, the economic exclusion zone (EEZ) of one or more states and the high seas". Sovereign responsibility must be worked out in collaboration with neighbouring coastal states and fishing entities. Usually this is done through the medium of an intergovernmental regional organisation set up for the purpose of coordinating the management of that stock.

Straddling stocks are usually pelagic, rather than demersal. Demersal species move less than pelagic species, since they tend to relate to bottom topography. Pelagic species are more mobile, their movements influenced by ocean temperatures and the availability of zooplankton as food. Example pelagic fish are capelin, herring, whiting, mackerel and redfish, There are, however, a few demersal species that are straddling, such as the Greenland halibut migrates in feeding/spawning migrations to Greenland in the west and to the Faeroes in the east.

Straddling stock can be compared with transboundary stock. Straddling stock range both within an EEZ as well as in the high seas. Transboundary stock range in the EEZs of at least two countries. A stock can be both transboundary and straddling.

## Fish Mortality

Fish mortality is a term widely used in fisheries science that denotes the loss of fish from a stock through death. The term is also commonly used in British English as a synonym for fish kill. Fish mortality can be divided into two types:

- Natural mortality: the removal of fish from the stock due to causes not associated with fishing. Such causes can include disease, competition, cannibalism, old age, predation, pollution or any other natural factor that causes the death of fish. In fisheries population dynamics natural mortality is denoted by (M).
- Fishing mortality: the removal of fish from the stock due to fishing activities using any fishing gear. It is denoted by (F) in fisheries models. (M) and (F) are additive instantaneous rates that sum up to (Z), the instantaneous total mortality coefficient; that is, Z=M+F. These rates are usually calculated on an annual basis.

Estimates of fish mortality rates are often included in mathematical yield models to predict yield levels obtained under various exploitation scenarios. These are used as resource management indices or in bioeconomic studies of fisheries.

## Natural

Estimating natural mortality (M) is one of the most difficult and critical elements of a stock assessment (Hewitt et al. 2007). There are two basic approaches used to estimate natural mortality: tagging studies and growth parameters. Tagging studies are used in the Brownies Model, where multiyear tagging studies are used to estimate natural mortality based on recaptures:

fi = ri(1 - Si)

The Paulys Model (using growth parameters) is an indirect way of estimating natural mortality. It assumes that there is a relationship between size and natural mortality. Pauly's original method was based on the correlation of M with von Bertalanffy growth parameters (K and L?) and temperature (Gunderson 2002):

N0 = N 1*e(-Z*?t)

In the Hoenigs method, M was inversely correlated with longevity across a wide variety of taxa (Hewitt and Hoenig 2005):

Ln(M) = 1.44-0.982*ln(tmax)

## Fishing

Fishing mortality (F) can be estimated by dividing the catch by the mean stock size. The catch includes annual commercial and recreational landings, along with dead discards. Bycatch discards would be estimated by estimating the percent of fish that are captured in a certain gear and the mortality associated with being captured in this gear. These mortality studies are generally done by using cages for a certain period of time following capture to determine the percentage of fish that die during being held in holding cages. These deaths are assumed to be associated with physical injury or physiological stress from being captured in the gear used during capture.

## Why estimating mortality is important

Mortality estimates are important to managers. Determining mortality rates are critical for determining abundance of fish populations. Using the model Z=M+F with M being Natural mortality and F being Fishing mortality (combined mortality from landings plus discard mortality) you can estimate the trend of a population. The mortality rates give you the total deaths of a population when you compare these to the total births or recruits to the population, you can determine if a population is increasing or decreasing. Knowing these rates can help managers to set harvest limits to (MSY) maximum sustainable yield or (OSY) optimum sustainable yield to give the maximum benefit to the stakeholders of the resource.

## Stock assessment

Stock assessments provide fisheries managers with the information that is used in the regulation of a fish stock. Biological and fisheries data are collected in a stock assessment.

A wide array of biological data may be collected for an assessment. These include details on the age structure of the stock, age at first spawning, fecundity, ratio of males to females in the stock, natural mortality (M), fishing mortality (F), growth rate of the fish, spawning behavior, critical habitats, migratory habits, food preferences, and an estimate of either the total population or total biomass of the stock.

The following data regarding fisheries activities is collected: the kinds of fisherman in the fishery, commercial versus recreational, and the gear that is used (longline, rod and reel, nets, etc), pounds of fish caught by each type of fisherman, the fishing effort each kind of fisherman expends, the age structure of the fish harvested by each group of fisherman, the ratio of males to females that are captured, how the fish are marketed, the value of the fish to the different fisherman groups, and the time and geographic location of the best catches. Also in the assessment, geographical boundaries of different stocks or populations are defined. From the combined biological and fisheries data, the current status and condition of the stock is defined and managers use this assessment to predict how in the future, stocks will respond to varying levels of fishing pressure. Ultimately managers want to reduce the levels of overfishing that occurs and restore stocks that have been overfished.

## Defining stock

In fisheries management, stock refers to a harvested or managed unit of a fish. Typically stocks are divided based on geographical location and not based on individual population. Spanish mackerel are distributed from Maine to the Yucatan Peninsula in Mexico. They are divided into two stocks, based on whether they migrate northward along the eastern United States coast or if they migrate into the Gulf of Mexico. Each stock of Spanish mackerel does not represent discrete populations. Stocks are not always composed of a single species. Stocks can be composed of multiple species due to their being harvested together or as a form of convenience for managers. An example of a multispecies stock is river herring. Alewives and blueback herring are labeled as river herring for management purposes due to their similar physical appearances and being harvested together. Individuals within a stock are subdivided into cohorts. A cohort is a group of fish born in the same year within a population or stock.

## Gathering data

Data used in stock assessments can be classified as fishery-dependent data or fishery-independent data. Fishery-dependent data is collected from the fishery itself, using both commercial and recreational sources. There are a variety of methods for obtaining fishery-dependent data. The most common approach is to use recorded landings. Landings are a record of the amount of fish sold and the numbers are typically reported in total weight. Another common mode for acquiring fishery-dependent data is through portside sampling of the catch of both recreational and commercial fisherman to obtain age and length information on the stock. Other less common methods for obtaining data is through the use of onboard observers, self-reporting, telephone surveys, and vessel-monitoring surveys.

Fishery-independent data is obtained in the absence of any fishing activity. The majority of this data is collected by state and federal agencies. A wide variety of methods and gear types are used to acquire fishery-independent data. Sampling equipment can include trawls, seines, acoustic and/or video surveys. The study may focus on a single species, multiple species, or a specific age range or cohort. Regardless of the method or approach, these surveys provide managers with an estimate of abundance. Mark and recapture studies are commonly used to estimate movement, migration, growth rate, natural mortality, and discard mortality. Stock assessments are often completed using both fishery-dependent and fishery-independent data.

## OVERFISHED VERSUS OVERFISHING

Overfished refers to the number of fish in the stock. Typically a stock is described as being overfished when current biomass of a stock is lower than what is required to support the maximum sustainable yield. Overfishing describes the rate of removal from a stock and can be categorized as two different types: recruitment and economic. Recruitment overfishing is when fishing pressure is too heavy to allow a fish population to replace itself. Economic overfishing occurs when the level of fish harvesting that is higher than that of economic efficiency and more fish are harvested than necessary to have maximum profits for the fishery.

### Biological reference points

Biological reference points are the primary output of stock assessments and fishing regulations are set to meet these biological benchmarks. There are several types of benchmarks are used depending on the preference of the regulating agency. The benchmarks are used to control different aspects of the fishery or population. Benchmarks that regulate fishing mortality include Fcurr, Fmsy, Emsy, and MFMT. Fcurr is the current level of fishing mortality, Fmsy is the fishing mortality that produces the maximum sustainable yield, and Emsy is the expected level of fishing that will produce the maximum sustainable yield. SSBcurr, SSBmsy, and MSST are used in the regulation of biomass levels. SSBcurr stands for the current spawning stock biomass, SSBmsy is the amount of spawning stock biomass needed to produce the maximum sustainable yield, and MSST is the minimum standing stock threshold. The amount that is available to be harvested is controlled by the benchmarks MSY and OY. MSY is the maximum sustainable yield and OY is the optimum yield of a stock. The main difference between MSY and OY is that MSY considers only the biology of the fish while the OY considers the economic aspect of the harvest.

Some of these benchmarks are combined in the form of ratios in order to better understand the status of the stock. A Fcurr/Fmsy ratio greater than one indicates that overfishing is occurring. When SSBcurr/SSBmsy is greater than one, the MSY will be produced, but if it is less than one it means that the stock is overfished. MFMT and MSST are used as limit reference points. When MFMT is exceeded or the spawning stock size dips below MSST the fishery is shut down. A benchmark that is currently gaining in popularity is SPR, spawning potential ratio. The SPR is the average fecundity of a recruit over its lifetime when the stock is fished divided by the average fecundity of a recruit over its lifetime when the stock is unfished. The SPR is based on the principle that certain levels of fish have to survive in order to spawn and replenish the stock at a sustainable level.

### Catch curve

A catch curve is a descriptive figure that describes catch by age and length. A catch curve only reflects fish that have recruited to a fishery and does not reflect the full age structure of a stock. A catch curve illustrates the proportions that different age and size classes are harvested by a fishery.

### Assessment models

The mathematical and statistical techniques used to complete a stock assessment are referred to as assessment models. Three commonly used models are surplus production models, statistical catch at age models, and virtual population analysis models. Of these models, surplus production models are the least complex and require the least amount of data. This model describes the stock solely in regards of biomass and the only used total catch and effort data. These are most commonly used in situations limited data is available on a stock. Statistical catch at age models are based on the age structure of a fished population. These models use the proportional catch-at-age to predict the relative abundance of each age class. These calculated relative abundances are then used to estimate future abundances of the stock and harvest regulations are set based on the predicted future abundances. In virtual population analysis models, catch-at-age data is used to estimate historical stock abundance. From this analysis, the manager then determines if overfishing is occurring. The type of model used depends on the data that is available. Modern stock assessment methods use statistical approaches to "integrate" multiple sources of information to estimate management quantities and their associated uncertainty.

### After a stock assessment

After a stock assessment is completed, the findings are provided to regional fishery management councils. The council identifies stocks that are endanger of or currently undergoing overfishing and then develops fishery management plans and regulations. Fishery management plans (FMP) must protect fishery resources while maintaining opportunities for domestic recreational and commercial fishing at sustainable levels of effort and yield. If a stock already has a FMP, the stock assessment is used to modify the FMP in response to current conditions. The public are also encouraged to participate in the management process and public hearings are held to allow for comment before the new management policies are enforced.

## FISH MEASUREMENT

Fish measurement refers to the measuring of the length of individual fish and of various parts of their anatomy. These data are used in many areas of ichthyology, including taxonomy and fisheries biology.

### Overall length

- Standard length (SL) refers to the length of a fish measured from the tip of the snout to the posterior end of the last vertebra or to the posterior end of the midlateral portion of the hypural plate. Simply put, this measurement excludes the length of the caudal fin.

- Total length (TL) refers to the length from the tip of the snout to the tip of the longer lobe of the caudal fin, usually measured with the lobes compressed along the midline. It is a straight-line measure, not measured over the curve of the body.

Standard length measurements are used with Teleostei (most bony fish), while total length measurements are used with Myxini (hagfish), Petromyzontiformes (lampreys), and (usually) Elasmobranchii (sharks and rays), as well as some other fishes.

In addition, fishery biologists often use a third measure, fork length (FL), in fishes with forked tails. This measure is the length from the tip of the snout to the end of the middle caudal fin rays and is used in fishes in which it is difficult to tell where the vertebral column ends.

### Other measurements

Other measurements that may be taken include the lengths of various fins, the lengths of fin bases, the length from the snout to various points on the body, and the diameter of the eye.

### Fish counter

Automatic fish counters are automatic devices for measuring the number of fish passing along a particular river in a particular period of time. Usually one particular species is of interest. One important species studied by fish counters are Atlantic salmon. This species is of interest owing to its ecologically vulnerable status and anadromous lifestyles.

## METHODS OF OPERATION

Fish counters can be divided into three principal types: resistive counters, optical counters, and hydroacoustic counters.

### Resistive counters

A resistive counter is associated with an in-river structure, such as a Crump weir. The resistivity of a fish is lower than that of water. So, as fish cross this barrier, they pass embedded electrodes, and the difference in resistivity disturbs the field established in the vicinity of the electrodes, altering inter-electrode resistance. With three electrodes these disturbances can then be measured by a Wheatstone bridge, or other means, to detect the size and direction of travel of the fish.

Fish counters of this type are used widely in Scotland to census populations of Atlantic salmon, where comparison with closed circuit television shows around a 97% detection rate.

## Optical counters

An optical counter is also associated with an in-river structure. However, rather than pass electrodes, in an optical counter the fish interrupt some of a number of vertically arranged beams of light. The pattern of beam-breaks can be used to determine the size, profile, and direction of motion of the fish.

The performance of optical counters has been determined by studies, under various conditions, to be greater than 90%. Optical counters can also distinguish the size of fish more accurately than other counter types and so are particularly useful where a mixture of species inhabit a river (for example rivers where salmon mix with sea trout).

The key disadvantage of optical counters is the small penetration of the beams through the water, restricting their use to narrow river features or in-river structures, such as fish ladders.

## Hydroacoustic counters

Hydroacoustic counters operate using the principles of sonar. A fish is insonified by a sound source, and reflections from the fish are detected by an underwater microphone. The reflection occurs because of the sudden change in impedance to sound waves within the fish, particularly at the swimbladder (90% of the reflection).

Hydroacoustic counters do not require in-river structures, but require skilled installation and operators. Without skilled installation at ideal sites hydroacoustic counters can be inaccurate. Studies typically indicate detection rates of 50% to 80%, though one study found detection rates as low as 3%. Careful planning and pre-siting study must be used to determine effectiveness.

The lack of a requirement for any in-river structure makes the counters an attractive proposition. Generally used for short-term or seasonal studies, some situations require a long-term count which is accurate in absolute terms, not only in relative change (for example, no hydroacoustic sensors are routinely used in the detection of Scottish Atlantic salmon). In these instances resistivity or optical sensors tend to be preferred. Such methods usually require significant habitat modification, such as construction of a weir to funnel the fish through the counter.

Recent advances in automated hydroacoustic monitoring systems has allowed continuous monitoring for periods exceeding 18 months. These systems include intelligent monitoring and real-time data processing, ensuring proper operation and publication of status and results (e.g. fish counts) on a routine basis.

# SITING COUNTERS

## In river structures

Resistivity and (particularly) optical fish counters require in-river structures to direct the fish through the detection aperture of the counter. Fish ladders and Borland fish passes are effective structures for this purpose and occasionally a natural restriction within the river may be used for a similar purpose. However for most counters a custom in-river structure will be required. One of the most effective such structures is the Crump weir, a triangular profile weir designed to ensure rapid planar flow over the detector.

## Siting within the river system

When monitoring anadromous fish such as the Atlantic salmon it is important to remember that a species may return to a particular breeding ground throughout its life. This means that within the larger rivers a number of quite distinct populations may cross a counter together, in aggregate. A population which uses a particular tributary may collapse whilst the overall numbers are not clearly affected. Issues with the management of that particular tributary and population therefore go unnoticed. It is important in these situations, therefore, that counters are placed to count individual populations, rather than the species in aggregate, in order that population collapses and recoveries can be detected.

## Data storage tag

A data storage tag (DST), also sometimes known as an archival tag, is a data logger that uses sensors to record data at predetermined intervals. Data storage tags usually have a large memory size and a long lifetime. Most archival tags are supported by batteries that allow the tag to record positions for several years. Alternatively some tags are solar powered and allow the scientist to set their own interval; this then allows data to be recorded for significantly longer than battery-only powered tags.

Data storage tags can have a variety of sensors; temperature, depth, light, salinity, pressure, pitch and roll, GPS, magnetic and compass. They can be used internally or externally in fish, marine animals or research animals. They are also used in other industries such as the food and beverage industry.

At the end of the monitoring period, the loggers can be connected to a computer and the data uploaded for analysis. Archival tags archive data to internal memory. Once they are recovered the data is then extracted by the researcher. The tag is generally mounted to the animal either by cutting a slit into the animal, inserting the tag, and sewing the opening closed. Alternatively researchers externally attach tags to animals by running anchor lines through the tag and into the dorsal fin for most "fish" species. For turtles the tag is epoxied to the shell of the turtle.

## Fisheries acoustics

Fisheries acoustics includes a range of research and practical application topics using acoustical devices as sensors in aquatic environments. Acoustical techniques can be applied to sensing aquatic animals, zooplankton, and physical and biological habitat characteristics.

Biomass estimation is a method of detecting and quantifying fish and other marine organisms using sonar technology. As the sound pulse travels through water it encounters objects that are of different density than the surrounding medium, such as fish, that reflect sound back toward the sound source. These echoes provide information on fish size, location, and abundance. The basic components of the scientific echo sounder hardware function is to transmit the sound, receive, filter and amplify, record, and analyze the echoes. While there are many manufacturers of commercially available "fish-finders," quantitative analysis requires that measurements be made with calibrated echo sounder equipment, having high signal-to-noise ratios.

### Acoustic tag

An acoustic tag is a small sound-emitting device that allows the detection and/or remote tracking of fish in three dimensions. Acoustic tags are commonly used to monitor the behavior of fish. Studies can be conducted in lakes, rivers, tributaries, estuaries or at sea. Acoustic tag tracking technology allows researchers to view 3D fish tracks in real-time with sub-meter resolution. Acoustic tags have been employed to help public utility agencies, private firms, and state and federal agencies meet fisheries regulations as defined by the Federal Regulations and Oversight of Energy known as FERC.

Acoustic tags come in an assortment of sizes and weights. To-date, the smallest acoustic tag successfully employed to provide fish behavior data in three dimensions was approximately 0.65 grams (tracking a juvenile salmonid at a length of 125 mm). Tag size increases significantly when tracking larger fish for longer periods of time (i.e. sturgeon).

## GIS AND AQUATIC SCIENCE

Geographic Information Systems (GIS) is becoming an integral part of aquatic science and limnology. Water by its very nature is dynamic. Features associated with water are thus ever-changing. To be able to keep up with these changes, technological advancements have given scientists methods to enhance all aspects of scientific investigation, from satellite tracking of wildlife to computer mapping of habitats. Agencies like the US Geological Survey, US Fish and Wildlife Service as well as other federal and state agencies are utilizing GIS to aid in their conservation efforts.

GIS is being used in multiple fields of aquatic science from limnology, hydrology, aquatic botany, stream ecology, oceanography and marine biology. Applications include using satellite imagery to identify, monitor and mitigate habitat loss. Imagery can also show the condition of inaccessible areas. Scientists can track movements and develop a strategy to locate locations of concern. GIS can be used to track invasive species, endangered species, and population changes.

One of the advantages of the system is the availability for the information to be shared and updated at any time through the use of web-based data collection.

## EcoSCOPE

The ecoSCOPE is an optical sensor system, deployed from a small remotely operated vehicle (ROV) or fibre optic cable, to investigate behavior and microdistribution of small organisms in the ocean. Although an ROV may be very small and quiet, it is impossible to approach feeding herring closer than 40 cm. The ecoSCOPE allows observation of feeding herring from a distance of only 4 cm. From 40 cm, the herrings' prey (copepods) in front of the herring are invisible due to the deflection of light by phytoplankton and microparticles in highly productive waters where herring live. With the ecoSCOPE, the predators are illuminated by natural light, the prey by a light sheet, projected via a second endoscope from strobed LEDs (2 ms, 100% relative intensity at 700 nm, 53% at 690 nm, 22% at 680 nm, 4% at 660 nm, 0% at 642 nm).

By imitating the long, thin snout of the garfish protruding into the security sphere of the alert herrings, an endoscope with a tip diameter of 11 mm is used. The endoscope is camouflaged to reduce the brightness-contrast against the background: the top is black and the sides are silvery. Additionally, the front of the ROV is covered by a mirror, reflecting a light gradient resembling the natural scene and making the instrument body virtually invisible to the animals. A second sensor images other copepods, phytoplankton and particles at very high magnification. Another advantage of these small "optical probes" is the minimal disruption of the current-field in the measuring volume, allowing for less disturbed surveys of microturbulence and shear.

Another video can be seen in the article for Atlantic herring.

An ecoSCOPE was also deployed to measure the dynamics of particles in a polluted estuary: see image on Particle (ecology), another as an underwater environmental monitoring system, utilizing the orientation capacity of juvenile glasseel.

## FishBase

FishBase is a comprehensive database of information about fish. As of May 2010, it included descriptions of 31,600 species, 279,100 common names in hundreds of languages, 49,300 pictures, and references to 44,200 works in the scientific literature.

In 1987, Daniel Pauly, inspired by the Species Identification Sheets and other products Walter Fischer had generated for the Food and Agriculture Organization in the 1970s, proposed a standardized database for fish species, as a part of the "ICLARM Software Project". The following year, he began to work with Rainer Froese, who had been working on an expert system to identify fish larvae. After a first attempt to build a system using Prolog, Froese switched to DataEase, a relational database for DOS. In 1989 the project received its first grant.

FishBase first appeared on the World Wide Web in August 1996, and a webmaster was hired in the following year. In 1999 a new module on ichthyoplankton and detailed data on fish larvae identification and rearing was created (LarvalBase) under the supervision of Bernd Ueberschaer. Eventually, the complete data of the CDs became available online.

As awareness of FishBase has grown among fish specialists, it has attracted over 1,700 contributors and collaborators. In order to preserve its value as a scientific database, FishBase is not allowed to include original data; all of its content must be based on previously-published material.

# 7

# ENVIRONMENTAL IMPACT OF FISH DISEASES

The environmental impact of fishing can be divided into issues that involve the availability of fish to be caught, such as overfishing, sustainable fisheries, and fisheries management; and issues that involve the impact of fishing on other elements of the environment, such as by-catch.

These conservation issues are part of marine conservation, and are addressed in fisheries science programs. There is a growing gap between how many fish are available to be caught and humanity's desire to catch them, a problem that gets worse as the world population grows.

Similar to other environmental issues, there can be conflict between the fishermen who depend on fishing for their livelihoods and fishery scientists who realise that if future fish populations are to be sustainable then some fisheries must reduce or even close.

The journal Science published a four-year study in November 2006, which predicted that, at prevailing trends, the world would run out of wild-caught seafood in 2048. The scientists stated that the decline was a result of overfishing, pollution and other environmental factors that were reducing the population of fisheries at the same time as their ecosystems were being degraded. Yet again the analysis has met criticism as being fundamentally flawed, and many fishery management officials, industry representatives and scientists challenge the findings, although the debate continues. Many countries, such as Tonga, the United States, Australia and New Zealand, and international management bodies have taken steps to appropriately manage marine resources.

## EFFECTS ON HABITAT

Some fishing techniques also may cause habitat destruction. Dynamite fishing and cyanide fishing, which are illegal in many places, harm surrounding habitat. Bottom trawling, the practice of pulling a fishing net along the sea bottom behind trawlers, removes around 5 to 25% of an area's seabed life on a single run. A 2005 report of the UN Millennium Project, commissioned by UN Secretary-General Kofi Annan, recommended the elimination of bottom trawling on the high seas by 2006 to protect seamounts and other ecologically sensitive habitats.

In mid October 2006, U.S. President Bush joined other world leaders calling for a moratorium on deep-sea trawling, a practice shown to often have harmful effects on sea habitat and, hence, on fish populations.

## Blast fishing

Blast fishing or dynamite fishing is the practice of using explosives to stun or kill schools of fish for easy collection. This often illegal practice can be extremely destructive to the surrounding ecosystem, as the explosion often destroys the underlying habitat (such as coral reefs) that supports the fish. The frequently improvised nature of the explosives used also means danger for the fishermen as well, with accidents and injuries.

Although outlawed, the practice remains widespread in Southeast Asia, as well as in the Aegean Sea and coastal Africa. In the Philippines, where the practice has been well documented, blast fishing was known prior to World War I, as this activity is mentioned by Ernst Jünger in his book Storm of Steel. One 1999 report estimated that some 70,000 fishermen (12% of the Philippines' total fishermen) engaged in the practice.

Extensive hard-to-patrol coastlines, the lure of lucrative, easy catches, and in some cases outright apathy or corruption on the part of local officials make enforcement of blast fishing bans an ongoing challenge for authorities.

Commercial dynamite or, more commonly, homemade bombs constructed using a glass bottle with layers of powdered potassium nitrate and pebbles or an ammonium nitrate and kerosene mixture are often employed. Such devices, though, may explode prematurely without warning, and have been known to injure or kill the person using them, or innocent bystanders.

Underwater shock waves produced by the explosion stun the fish and cause their swim bladders to rupture. This rupturing causes an abrupt loss of buoyancy; a small number of fish float to the surface, but most sink to the sea floor. The explosions indiscriminately kill large numbers of fish and other marine organisms in the vicinity and can damage or destroy the physical environment, including extensive damage to coral reefs.

## Strategies to control blast fishing

### *Community-based Enforcement*

In Tanzania, one of the few methods to help manage blast fishing is a joint approach between fisheries officers and village committees. Working together, they help the enforcement agencies recognize offenders by patrolling the sea as well as providing information collected in the local villages. As a result, this has assisted the enforcement agencies to reduce the occurrence of fish blasting from an average of 8 per day to zero. It has also provided sustainable funding to continue the efficient patrols, a certified planning institution, and suitable training and information to prosecutors and judges.

Similar patrols employed in Indonesia and Philippines have reduced the amount of blast fishing occurrences there. Based on dialogue with stakeholder groups in Southeast Asia and people of Tanzania and Philippines, it is evident that firmer enforcement is an effective strategy in managing blast fishing. Many countries have laws regarding blast fishing, but they are not fully implemented. Effective management of Marine Protected Areas (MPAs) is key in the patrolling of illegal fishing areas.

### *Blast Detection System*

This method involves a triangulation system of hydrophones one meter apart that is capable of detecting blast events and at the same time eliminating other sources of underwater noise. The goal of the system is to improve and assist the effectiveness of fisheries patrol. Based on tests performed in Malaysia from 7 to 15 July 2002, a total of 13 blasts were recorded with a directional uncertainty of 0.2°. An electronic compass would limit the bearing uncertainty to 0.2° while correcting for the local magnetic effects of ferrous metals, therefore making sure the precision of the system is high.

Similar triangulation systems of hydrophones can potentially locate single blast events within 30 m at a range of 10 km. The detector system can be mounted on a patrol boat to help locate a probable range of blasts. Two or more patrol boats would permit accurate triangulation of blast events. Such a method is also beneficial enforcement agencies, as it offers stronger evidence to support convictions related to blast fishing.

## Countries

### *Tanzania*

In northern Tanzania, blast fishing, which is illegal, has re-emerged in recent years as a key danger to its coral reefs. This has occurred even though major institutions like local communities and the district government have been put in place for enhanced fisheries management. The damage of blast fishing in the area has contributed to unstable coral reefs, discouragement of tourism investors, and a threat to the habitat of coelacanths in the region. Other impacts of blast fishing in the area include reports that citizens have died or lost limbs due to the blasting. The northern part of the country has many beautiful beaches and uninhabited islands. However, many investors feel and tourists are discouraged due to the fish blasting.

In Tanzania, coral reefs are essential for both ecological and socio-economic reasons. They are full of fish, lobsters, prawns, crabs, octopuses, mollusks, and sea cucumbers. In addition, coral reefs are one of the major tourist attractions in Tanzania. The coastal tourism provides a living for the people as well as foreign currency for the country. However, there has been an increase in the people living along the coast which has led to a large demand for fisheries. It has led to overexploitation and destructive fishing practices. Blast fishing has been practised in Tanzania since the 1960s. It was during the 1980s and 1990s that blast fishing was at its peak in Tanzania. For example, in Mnazi bay, Mtwara, 441 blasts were recorded in two months in 1996, and 100 blasts were witnessed through one 6 hour period in Mpovi reef.

***Indonesia***

Blast fishing in Indonesia has been around for over 50 years. The use of bombs made with kerosene and fertilizer is very popular in the region. In the market in the city of Makassar, an estimated 10 to 40 percent of the fish are caught in this manner. The local fishermen find the technique to be easier and more productive than traditional methods. The goal for the country has been to implement stricter polices and fisheries management programs to limit the killing of the fish as well as the destruction to the marine ecosystem. Forty years ago, blast fishing was practiced with dynamite which was in plentiful supply after World War II. Today, fishermen mostly use homemade bombs that are made from bottles filled with an explosive mixture; weights are also added to make the bottle sink faster underwater. After the bomb explodes, the fish killed or stunned by the shock wave from the explosion are collected.

## Cyanide Fishing

Cyanide fishing is a method of collecting live fish mainly for use in aquariums, which involves spraying a sodium cyanide mixture into the desired fish's habitat in order to stun the fish. The practice hurts not only the target population, but also many other marine organisms, including coral and thus coral reefs. Cyanide fishing is practiced mainly in saltwater fishing regions of Southeast Asia. Since the practice was never widely publicized or officially approved, its origins are uncertain, but it is believed to have originated in the 1950s in the Philippines. Later in the 20th century the practice was adopted by some fishing outfits in Indonesia, Thailand, Maldives, and Taiwan, among others. Cyanide fishing was initially developed to stun and capture fish for aquariums and collectors, but it was soon used for catching food fish as well. It is illegal in many of the countries in which it is practiced, although these laws are often minimally enforced. Grouper, wrasse, and coral trout are among the more popular species of fish captured through cyanide fishing.

The World Resources Institute determined that approximately 20% of the live fish traded on the Philippine market in 1996 were caught using cyanide; assuming this is reflective of southeast Asian practice as a whole, environmental engineer David Dzombak estimates that 12,000 to 14,000 tons of live food fish are caught each year using this method.

Colourful, particularly eccentric, and therefore rare coral fish are packed into plastic bags; up to two thirds of these fish die during transport. Estimates suggest 70 to 90% of aquarium fish exported from the Philippines are caught with cyanide. Due to the post-capture handling stress and the effects of the cyanide, fish are bound to have a shorter life span than usual in aquariums. According to an interview with experienced aquarium owners, they were willing to pay more for net-caught fish because of the higher survival rate. They also said they would not trust an eco-labelling system, which can be misleading.

The basis for this illegal fishing method is, among others, the rising demand for live fish in the higher-class restaurants of the big cities, particularly in rich, nearby countries, which pay increasingly high prices. The extremely low wages of the fishermen in remote, underdeveloped areas, where there are no alternative sources of income, drive them to endure the health risks and possible prosecution.

## Method

The fishermen dive into the sea usually without artificial breathing aids, although some use illegal and highly-dangerous apparatus (commonly garden hose surface-fed from the type of air compressor commonly used to power jackhammers). When they reach the coral reefs, they spray the poison between the individual layers, after which the yield is collected. Edible fish, of which a number are sold for general consumption, are first placed for ten to fourteen days in fresh water for "rinsing". Recent studies have shown that the combination of cyanide use and stress of post capture handling results in mortality of up to 75% of the organisms within less than 48 hours of capture. With such high mortality numbers, a greater number of fish must be caught in order to supplement post catch death.

Cases have been reported of fishermen dumping drums of concentrated cyanide in places where fishing is difficult or economic times are hard. Such high concentrations normally kill most of the haul, but in these cases the objective is no longer to catch live fish, but to catch the largest amount period.

## Mechanism

In seawater sodium cyanide breaks down into sodium and cyanide ions. In humans, cyanides block the oxygen-transporting protein haemoglobin; the haemoglobin in fish is closely related to that of humans, and can combine with oxygen even faster. Through the irreversible combining of cyanide ions onto the active structural domain, oxygen is prevented from reaching the cells, and an effect similar to carbon monoxide poisoning results. Coral polyps, young fish and spawn are most vulnerable; adult fish can take somewhat higher doses. The use of cyanide is known to cause mortality on laboratory corals in measured doses, however these data are very difficult to quantify in regard to wild populations. In humans ingestion or breathing in of cyanide leads to unconsciousness within a minute; asphyxiation follows. Lower doses lead to temporary or permanent disability and/or sensory failure. This is a constant danger for the fishermen; there are many local accounts of such "occupational accidents", but such incidents are not recorded in official statistics or statements.

## Habitat Destruction

Many fishing and diving areas across the whole of South East Asia, already severely damaged from the impact of dynamite fishing, have been ruined or totally lost through cyanide fishing. Cyanide concentration slows photosynthesis in zooxanthellae, which results in coral reefs losing color; it also eliminates one of their major food sources. Even at very low doses, cyanide results in higher mortality levels among corals.

Most legal and illegal fishing methods cannot by themselves destroy a stable ecosystem. However, through the effects of synergy, they have led to the breakdown of large coastal areas which were formerly robust fishing grounds.

## OVERFISHING

Overfishing has also been widely reported due to increases in the volume of fishing hauls to feed a quickly growing number of consumers. This has led to the breakdown of some sea ecosystems and several fishing industries whose catch has been greatly diminished. The extinction of many species has also been reported. According to an FAO estimate, over 70% of the world's fish species are either fully exploited or depleted. According to Nitin Desai, Secretary General of the 2002 World Summit on Sustainable Development, "Overfishing cannot continue, the depletion of fisheries poses a major threat to the food supply of millions of people."

The cover story of the May 15, 2003 issue of the science journal Nature - with Dr. Ransom A. Myers, an internationally prominent fisheries biologist (Dalhousie University, Halifax, Canada) as the lead author - was devoted to a summary of the scientific information. The story asserted that, as compared with 1950 levels, only a remnant (in some instances, as little as 10%) of all large ocean-fish stocks are left in the seas. These large ocean fish are the species at the top of the food chains (e.g., tuna, cod, among others). However, this article was subsequently criticized as being fundamentally flawed, although much debate still exists (Walters 2003; Hampton et al. 2005; Maunder et al. 2006; Polacheck 2006;Sibert et al. 2006) and the majority of fisheries scientists now consider the results irrelevant with respect to large pelagics.

### Types

There are three recognized types of overfishing: growth overfishing, recruit overfishing and ecosystem overfishing.

Growth overfishing - is when fish are harvested at an average size that is smaller than the size that would produce the maximum yield per recruit. This makes the total yield less than it would be if the fish were allowed to grow to a reasonable size. It can be countered by reducing fishing mortality to lower levels and increasing the average size of the fish harvested to a length that will allow maximum yield per recruit.

Recruit overfishing - is when the mature adult (spawning biomass) population is depleted to a level where it no longer has the reproductive capacity to replenish itself. There are not enough adults to produce offspring. Increasing the spawning stock biomass to a target level is the approach taken by managers to restore an overfished population to sustainable levels. This is generally accomplished by placing moratoriums, quotas and minimum size limits on a fish population.

Ecosystem overfishing - is when the balance of the ecosystem is altered due to overfishing. Declines in the abundances of large predatory species declines and in turn small forage type species increase in abundance, causing a shift in the balance of the ecosystem towards smaller species of fish.

## Instances

Examples of the outcomes from overfishing exist in areas such as the North Sea of Europe, the Grand Banks of North America and the East China Sea of Asia. In these locations, overfishing has not only proved disastrous to fish stocks but also to the fishing communities relying on the harvest. Like other extractive industries such as forestry and hunting, fishery is susceptible to economic interaction between ownership or stewardship and sustainability, otherwise known as the tragedy of the commons.

The Peruvian coastal anchovy fisheries crashed in the 1970s after overfishing, following an El Niño season which largely depleted anchovies from its waters. Anchovies had previously been a major natural resource in Peru; indeed, 1971 alone yielded 10.2 million metric tons of anchovies. However, in the following year, and the four after that, the Peruvian fleet's catch amounted to only about 4 million tons. This was a major loss to Peru's economy.

- The collapse of the cod fishery off Newfoundland, and the 1992 decision by Canada to impose an indefinite moratorium on the Grand Banks, is a dramatic example of the consequences of overfishing.
- The sole fisheries in the Irish Sea, the west English Channel, and other locations have become overfished to the point of virtual collapse, according to the UK government's official Biodiversity Action Plan. The United Kingdom has created elements within this plan to attempt to restore this fishery, but the expanding global human population and the expanding demand for fish has reached a point where demand for food threatens the stability of these fisheries, if not the species' survival.
- Many deep sea fish are at risk, such as orange roughy, Patagonian toothfish and sablefish. The deep sea is almost completely dark, near freezing and has little food. Deep sea fish grow slowly because of limited food, have slow metabolisms, low reproductive rates, and many don't reach breeding maturity for 30 to 40 years. A fillet of orange roughy at the store is probably at least 50 years old. Most deep sea fish are in international waters, where there are no legal protections. Most of these fish are caught by deep trawlers near seamounts, where they congregate because of food. Flash freezing allows the trawlers to work for days at a time, and modern fishfinders target the fish with ease.

## Consequences

According to a 2008 UN report, the world's fishing fleets are losing $50 billion USD each year through depleted stocks and poor fisheries management. The report, produced jointly by the World Bank and the UN Food and Agriculture Organization

(FAO), asserts that half the world's fishing fleet could be scrapped with no change in catch. In addition, the biomass of global fish stocks have been allowed to run down to the point where it is no longer possible to catch the amount of fish that could be caught. Increased incidence of schistosomiasis in Africa has been linked to declines of fish species that eat the snails carrying the disease-causing parasites. Massive growth of jellyfish populations threaten fish stocks, as they compete with fish for food, eat fish eggs, and poison or swarm fish, and can survive in oxygen depleted environments where fish cannot; they wreak massive havoc on commercial fisheries. Overfishing eliminates a major jellyfish competitor and predator exacerbating the jellyfish population explosion.

## Acceptable levels

The notion of overfishing hinges on what is meant by an acceptable level of fishing. More precise biological and bioeconomic terms define acceptable level as follows:

- Biological overfishing occurs when fishing mortality has reached a level where the stock biomass has negative marginal growth (slowing down biomass growth), as indicated by the red area in the figure. (Fish are being taken out of the water so quickly that the replenishment of stock by breeding slows down. If the replenishment continues to slow down for long enough, replenishment will go into reverse and the population will decrease.)
- Economic or bioeconomic overfishing additionally considers the cost of fishing when determining acceptable catches. Under this framework a fishery is considered to be overfished when catches exceed maximum economic yield where resource rent is at its maximum. Fish are being removed from the fishery so quickly that the profitability of the fishery is sub-optimal. A more dynamic definition of economic overfishing also considers the present value of the fishery using a relevant discount rate to maximise the flow of resource rent over all future catches.

### *Harvest control rule*

A current model for predicting acceptable levels is the Harvest Control Rule (HCR), which is a variable over which management has some direct control as a function of some indicator of stock status. Constant catch and constant fishing mortality are two types of simple harvest control rules.

### *Input and output orientations*

Fishing capacity can also be defined following an input or an output orientation.

- An input-oriented fishing capacity is defined as the maximum available capital stock in a fishery that is fully utilized at the maximum technical efficiency in a given time period, given resource and market conditions.
- An output-oriented fishing capacity is defined as the maximum catch a vessel (fleet) can produce if inputs are fully utilized given the biomass, the fixed inputs, the age structure of the fish stock, and the present stage of technology.

Technical efficiency of each vessel of the fleet is assumed necessary to attain this maximum catch. The degree of capacity utilization results from the comparison of the actual level of output (input) and the capacity output (input) of a vessel or a fleet.

## Mitigation

With present and forecast levels of the world population it is not possible to solve the overfishing issue; however, there are mitigation measures that can save selected fisheries and forestall the collapse of others.

In order to meet the problems of overfishing, a precautionary approach and Harvest Control Rule (HCR) management principles have been introduced in the main fisheries around the world. The Traffic Light colour convention introduces sets of rules based on predefined critical values, which could be adjusted as more information is gained. The "United Nations Convention on the Law of the Sea" treaty deals with aspects of overfishing in articles 61, 62, and 65.

- Article 61 requires all coastal states to ensure that the maintenance of living resources in their exclusive economic zones is not endangered by over-exploitation. The same article addresses the maintenance or restoration of populations of species above levels at which their reproduction may become seriously threatened.
- Article 62 provides that coastal states: "shall promote the objective of optimum utilization of the living resources in the exclusive economic zone without prejudice to Article 61"
- Article 65 provides generally for the rights of, inter alia, coastal states to prohibit, limit, or regulate the exploitation of marine mammals.

Overfishing can be viewed as a case of the tragedy of the commons; in that sense, solutions would promote property rights, such as privatization and fish farming. Daniel K. Benjamin, in Fisheries are Classic Example of the "Tragedy of the Commons", cites research by Grafton, Squires, and Fox to support the idea that privatization can solve the overfishing problem:

According to recent research on the British Columbia halibut fishery, where the commons has been at least partly privatized, substantial ecological and economic benefits have resulted. There is less damage to fish stocks, the fishing is safer, and fewer resources are needed to achieve a given harvest.

Another possible solution, at least for some areas, is fishing quotas, so fishermen can only legally take a certain amount of fish. A more radical possibility is declaring certain areas of the sea "no-go zones" and make fishing there strictly illegal, so the fish in that area have time to recover and repopulate.

Controlling consumer behavior and demand is a key in mitigating action. Worldwide a number of initiatives emerged to provide consumers with information regarding the conservation status of the seafood available to them. The Guide to Good Fish Guides lists a number of these.

## Fishing Quotas

A model of the interaction between fish and fishers showed that when an area is closed to fishers, but there are no catch regulations such as individual transferable quotas, fish catches are temporarily increased but overall fish biomass is reduced, resulting in the opposite outcome than the one desired for fisheries. Thus, a displacement of the fleet from one locality to another will generally have little effect if the same quota is taken. As a result, management measures such as temporary closures or establishing a Marine Protected Area of fishing areas are ineffective when not combined with individual fishing quotas. An inherent problem with quotas is that fish populations vary from year to year. A study has found that fish populations rise dramatically after stormy years due to more nutrients reaching the surface and therefore greater primary production. To fish sustainably quotas need to be changed each year to take account of the population of fish but this is difficult to do.

## Individual Transferable Puotas

Individual transferable quotas (ITQs) are fishery rationalization instruments defined under the Magnuson-Stevens Fishery Conservation and Management Act as limited access permits to harvest quantities of fish. Fisheries scientists decide the optimal amount of fish (total allowable catch) to be harvested in a certain fishery, taking into account carrying capacity, regeneration rates and future values. Under ITQs, members of a fishery are granted rights to a percentage of the total allowable catch which can be harvested each year. These quotas can be fished, bought, sold, or leased allowing for the least cost vessels to be used. ITQs are used in New Zealand, Australia, Iceland, Canada and the United States. Only three ITQ programs have been implemented in the United States due to a moratorium supported by Ted Stevens.

In 2008 a large scale study of fisheries that used ITQ's and ones that didn't provided strong evidence that ITQ's can help to prevent collapses and restore fisheries that appear to be in decline.

### *Fishing Suspension*

China bans fishing in the South China Sea for a period each year.

### *Benefits of Underfishing*

Deliberately underfishing to increase long term fish stocks has been proposed as a way fisherman can maximize their yields in the long run.

### *Resistance from Fishermen*

There is always disagreement between fishermen and government scientists... Imagine an overfished area of the sea in the shape of a hockey field with nets at either end. The few fish left therein would gather around the goals because fish like structured habitats. Scientists would survey the entire field, make lots of unsuccessful hauls, and conclude that it contains few fish. The fishermen would make a beeline to the goals, catch the fish around them, and say the scientists do not know what they are talking about. The subjective impression the fishermen get is always that there's lots of fish - because they only go to places that still have them... fisheries scientists survey and compare entire areas, not only the productive fishing spots. - Fisheries scientist Daniel Pauly

- The fishing capacity problem is not only related to the conservation of fish stocks but also to the sustainability of fishing activity. Causes of the fishing problem can be found in property rights regime of fishing resources. Overexploitation and rent dissipation of fishermen arise in open-access fisheries as was shown in Gordon.
- In open-access resources like fish stocks, in the absence of a system like individual transferable quotas, the impossibility of excluding others provokes the fishermen who want to increase catch to do so effectively by taking someone else' share, intensifying competition. This tragedy of the commons provokes a capitalization process that leads them to increase their costs until they are equal to their revenue, dissipating their rent completely.

### *Removal of subsidies*

Several scientists have called for an end to subsidies paid to deep sea fisheries. In international waters beyond the 200 nautical mile exclusive economic zones of coastal countries, many fisheries are unregulated, and fishing fleets plunder the depths with state of the art technology. In a few hours, massive nets weighing up to 15 tons, dragged along the bottom by deep-water trawlers, can destroy deep-sea corals and sponge beds that have taken centuries or millennia to grow. The trawlers can target orange roughy, grenadiers or sharks. These fish are usually long-lived and late maturing, and their populations take decades, even centuries to recover.

Fisheries scientist Daniel Pauly and economist Ussif Rashid Sumaila have examined subsidies paid to bottom trawl fleets around the world. They found that $152 million US per year are paid to deep-sea fisheries. Without these subsidies, global deep-sea fisheries would operate at a loss of $50 million a year. A great deal of the subsidies paid to deep-sea trawlers is to subsidize the large amount of fuel required to travel beyond the 200 mile limit and drag weighted nets.

- "There is surely a better way for governments to spend money than by paying subsidies to a fleet that burns 1.1 billion litres of fuel annually to maintain paltry catches of old growth fish from highly vulnerable stocks, while destroying their habitat in the process" - Pauly.
- "Eliminating global subsidies would render these fleets economically unviable and would relieve tremendous pressure on over-fishing and vulnerable deep-sea ecosystems" - Sumaila.

### *Ecological disruption*

Fishing may disrupt food webs by targeting specific, in-demand species. There might be too much fishing of prey species such as sardines and anchovies, thus reducing the food supply for the predators. It may also cause the increase of prey species when the target fishes are predator species such as salmon and tuna. Fisheries can reduce fish stocks that cetaceans rely on for food.

# 8

# DIAGNOSING OF FISH DISEASES

It is frustrating when our fish are not healthy, however, there are things we can do to help them. some of the more commonly seen symptoms of illness in freshwater fish kept in aquariums.

When your fish aren't feeling well, the first thing you should do is check your water quality and take steps to resolve any water quality problems. Good water quality is the most important thing you can do to help keep your fish healthy. Second, try to identify specifically what the illness is based on symptoms. Third, choose a treatment based on your diagnosis and understand how that treatment works.

To be prepared for illnesses, make sure you have all the the necessary water test kits ready. You may also want to stock your medicine cabinet so that you have the things you need when your aquatic friends get sick. If you are unfamiliar with fish anatomy, make sure to check our external and internal anatomy articles. When treating with medications, you should be up to speed on your fish, as some species of fish don't tolerate certain medications well (e.g. Scaleless fish and salt, tetras and Formalin/Malachite Green).

To further familiarize yourself with medications read GAB Articles on Medications and Medicated Food. As you can see when you visit our GAB Sick Fish Forum, we believe in assessing the situation and finding the most appropriate solution when faced with fish problems. You may also want to check if there is an aquatic vet near you.

When you medicate, you usually want to follow the directions on your medication, turn off UV sterilizers, and remove carbon from your filter before treating with waterborne medications. It is also good to do a partial water change with a thorough vacuuming prior to beginning treatment. Depending on the situation, it may be best to treat in a hospital tank. Also remember that you can try to minimze the possibility of diseases by buying specimens that look healthy, quarantining them and feeding them good nutrious food.

## FLASHING/RUBBING

Flashing means that the fish is throwing itself against or rubbing against decorations, gravel or tank walls. It is a sign of discomfort and irritation. This irritation is likely felt externally so you are looking for external answers. Poor water quality or parasites is the usual explanation.

### Causes and Remedies

#### *Water Quality*

If your fish is flashing, this may be due to poor water quality. Check your water parameters and perform water changes if necessary. These tests are best done with your own testers. As a minimum, you should test for ammonia, nitrite, nitrate, and pH.

#### *Irritants in the Water*

In addition to ammonia and nitrite, other irritants in the water may cause flashing such as chlorine, metals, household cleaners, etc. Make sure you use a water conditioner that removes chlorine and binds heavy metals. If you suspect other chemicals, preform a large water changes and add fresh carbon to your filter. Also, if you have added decorations to your aquarium that are not aquarium safe, they may leach chemicals into your water. Remove the decoration, perform a large waterchange and add fresh carbon to your filter.

#### *Parasites*

Flashing may be a sign of external parasites, check the fish carefully for other symptoms such as slimecoat changes, spots or anchor worms. After you rule out water quality and the fish continue flashing, consider treating the tank for parasites with a product containing formalin and malachite green (e.g. Rid Ich or Quick Cure). If it is a goldfish or other salt tolerant fish, you can treat with salt as described here, however, there are salt-resistant species of costia that may not clear with salt. Goldfish are susceptible to flukes. If you have a microscope, you may want to scrape and scope to determine whether flukes are present, or if that is not possible, treat for flukes with praziquantel.

## GASPING

Gasping is when the fish are at the water surface trying to get air. This is also called piping. As with most symptoms, it is either the environment (water quality) or an illness that is causing this behavior. Piping is rather serious and you should try to identify the causes and remedy this immediately.

### *Causes and Remedies*

#### *Water Quality*

If your fish are gasping, this may be due to poor water quality. Check your water parameters and perform water changes if necessary. These tests are done best with your own testers. As a minimum you need to test for ammonia, nitrite, nitrate, and pH.

### *Oxygen Levels*

If you have ruled out water quality and fish are still gasping, they may not be getting enough oxygen. Add an air stone or venturi to increase surface agitation. Vacuum your gravel to remove debris (decaying debris consumes oxygen). Lower your water level so that your filter output breaks the water surface.

### *Nitrite Poisoning or Brown Blood Disease*

If your nitrite is not 0 ppm, look at the gills. They should be bright blood red and well defined. If they are brown, this indicates brown blood disease from nitrite poisoning. Nitrite keeps red blood cells from being able to transport oxygen. Perform water changes to lower nitrite. If the fish is salt tolerant, add some salt (1 tsp per gallon) to help protect against nitrite entering via the gills. The chloride ions in salt result in the fish absorbing less nitrite trough the gills. Methylene blue can be used to unbind the red blood cells. You can read more about nitrite poisoning here.

### *Chlorine or Chloramine*

Chlorine or chloramine can also cause gasping if you did not dechlorinate your water (or if your city added more chlorine/chloramine than your water conditioner could detoxify). As with nitrite poisoning, the gills will look brown instead of red. Methylene blue can be used to unbind the red blood cells. If you suspect high chlorine or chloramine levels, you could call your city's water quality department to find out how much disinfectant they add to your water supply. If this is the cause of the problem, you need to increase the amount of dechlorinator you use.

### *Poisons*

Review the latest events and consider whether something poisonous or toxic may have gotten into the tank (e.g. sprays, soap, etc.). If that is the case, do a large water change and add fresh carbon.

Sometimes the gas H2S (hydrogen sulfide) may leach out of the substrate. Hydrogen sulfide gas is formed by anaerobic bacterial action. If you (or the fish) disturb the substrate, this gas may bubble out and cause gasping. In severe cases, H2S poisoning can cause swirling, flip-over and death.

However, the gas is not that soluble in water, so the fish have to get a face-full to make them sick, especially if you have plenty of surface agitation in the tank. You should be able to identify this gas by its smell (much like rotten eggs). It is advisable to move your fish to another container with their tank water while you remove all but an inch of gravel and vacuum the tank well before returning the fish. To avoid this problem, vacuum the substrate regularly, or if you have heavily planted tanks be careful not to disturb the substrate too much.

### *Flukes or Other Gill Parasites*

Gill flukes and other parasites or bacterial infections can damage gills. Check the fish for symptoms of bacterial or parasitic problems and treat appropriately (praziquantel for gill flukes, formalin/malachite green for other parasites or antibiotics for bacterial infections).

## LETHARGY

If you notice that the fish is not behaving normally, but instead is hanging around the top of the tank or near the bottom and is not swimming around much, the fish is probably lethargic. This is a behavioral symptom that is a sure sign that something is wrong. Once you observe that a fish has become lethargic and after you've checked your water parameters, you have to look for other symptoms since lethargy is not specific enough to provide a definitive diagnosis.

### Causes and Remedies

#### *Water Quality*

If your fish is lethargic, this may be due to poor water quality. Check your water parameters and perform water changes if necessary. These tests are best done with your own liquid test kits. As a minimum you need to test for ammonia, nitrite, nitrate, and pH.

#### *Temperature Problem*

If the fish is not moving around much, it may be because the water is either too warm or too cold. Take care to measure the temperature. Tropical fish like a temperature around 78-79 F. For many tropical fish anything lower than 77 F may slow them and cause them to become stressed. Goldfish will do well in temperatures from 68 to 79 F. Most aquarium fish can tolerate temperatures up to 82-84 F, if you provide plenty of airation (warmer water holds less oxygen). To minimize stress, change the temperature of the water in your tank slowly/gradually.

#### *Infection*

If you fish is lethargic, check for other signs like redness, fuzz or fin deterioration that would suggest bacterial causes. If you cannot find any such signs (and you have ruled out water quality, parasites, egglaying, or dropping fry), check to see if the fish is still eating. If it is lethargic and not eating, it is probably an internal infection and it would be prudent to move the fish to a hospital tank. Feed antibiotic food (like Medi-Gold) or if the fish is not eating, treat with a broad spectrum in-tank antibiotic like kanamycin. If the fish is big enough you should consider injections. Keep checking for other signs of illness.

#### *Parasites*

Parasites can cause lethargy - particularly parasites that attack gills such as flukes or costia. Check carefully for signs of these parasites by looking at the slimecoat and seeing if the fish is gasping or flashing. Also if possible check the gills.

Lethargy may also be a sign of internal parasites. These are hard to diagnose as they do not leave many clear symptoms except lethargy, failure to thrive and losing weight despite eating. If you have access to a microscope, you can try to identify potential pathogens by looking at some poop emulsified in tank water. Treat for internal parasites with Jungle Anti-Parasite Medicated Fish Food.

## YAWNING

Sometimes we can observe fish making weird shapes with their mouths, namely forming their mouths into an 'o' shape and looking like they are yawning. An occasional 'yawn' should be of no concern, but if the fish does this repeatedly through the day there may be something bothering it. Often times fish display irritation in the gills by opening and closing the mouth rapidly.

The yawning (or mouth opening and closing) is very likely an attempt to clear the gills, therefore, those are the areas we focus on when we look for other symptoms. Try to check the gills of the fish to see it they are normal for the type of fish you have.

### Causes and Treatment

#### *Water Quality*

If your fish is yawning, this may be due to poor water quality. Check your water parameters and perform water changes if necessary. These tests are best done with your own testers. As a minimum you need to test for ammonia, nitrite, nitrate, and pH.

Yawning could also be caused by low oxygen levels, so add an air stone or venturi to increase surface agitation. Vacuum your gravel to remove debris (decaying debris consumes oxygen). Lower your water level so that your filter output breaks the water surface.

#### *Gill Parasites*

Several parasites can attack gills and cause stress to the fish.

- Flukes
- Protozoan parasites (ich, costia)
- Hexamita

To determine which parasite is involved, you need to look for other signs in addition to yawning. Is the fish lethargic, can you see spots or changes in the slime coat?

#### *Flukes*

Flukes (Dactylogyrus or Gyrodactylus) are parasites common in goldfish in particular. They cannot be seen with the naked eye, so you need a gill scrape or a body scrape to diagnose flukes. Here is a video of moving flukes from Koivet, and more information on flukes can be found here.

However, since scraping and scooping are not always possible, we sometimes treat for flukes if several of the symptoms are present and we have no other good explanation for the symptoms. Thankfully a very benign product, praziquantel, is now widely available. If it is not available locally, you can order it (in the form of Prazi-Pro) from outlets such as the Goldfish Connection, PondRX or MOPS (in Canada).

### *Protozoan Parasites*

Protozoan parasites (such as ich or costia) should also make the fish display other symptoms (flashing, slimecoat changes, spots) so look for those. These parasites are best treated with a malachite Green/formalin combination (such as in Rid Ich or Quick Cure), or if the fish is salt tolerant, plain salt.

Note: some strains of costia and ich are salt resistant. Also make sure you read the label of the medication to make sure your fish can tolerate it.

### *Hexamita*

If doing any of the above did not help, treat the fish for Hexamita with metrodonidazole (aka Flagyl). Or if this is not available, use a parasite medication that has a shot gun approach like Jungle Parasite Guard.

### *Other Gill Problems*

There may be other gill issues

# SPITTING OUT FOOD

If your fish does not eat well or spits out the food, there is very likely a problem. However, there are several possibilities for what this can be so you have to observe carefully when this happens.

## Causes and Treatment

### *Water Quality*

The behaviour may mean your fish is not feeling well, and water quality may be the under laying cause so check your water parameters and perform water changes if necessary. These tests are best done with your own testers. As a minimum you need to test for ammonia, nitrite, nitrate, and pH.

### *Fish does not like the food*

Some fish are very picky eaters. They will not take well to new offerings and can spit things out that feel different to them. Bettas for instances are notorious for their food preferences. Make sure that you are feeding food that fills the requirement for your species and give the fish time to adjust.

### *Problems with the mouth*

In some cases, spitting out food can mean that the mouth is sore. Observe carefully and if possible have a look inside the mouth. Some larger fish, like goldfish, may get gravel stuck in the mouth. If the fish cannot get rid of the gravel, you may have to try to remove it with tweezers. Another possibility is mouth rot - a form of the illness called flex or columnaris caused by the bacterium Flavobacterium columnaris.

### *Gill Parasites*

Several parasites can attack gills and cause stress to the fish

- Flukes
- Protozoan parasites (ich, costia)
- Hexamita

To determine which parasite is involved, you need to look for other signs as well. Is the fish lethargic, can you see spots or changes in the slime coat?

### *Illness or Infection*

Sometimes a fish can be too lethargic and sick to eat because it is fighting another illness. Look for other symptoms to the help you diagnose this illness.

### *Mouth Brooder*

Some species are mouth brooders and will carry their fry in the mouth. During that period, eating will not be the priority. This is natural part of the fish's life and nothing to worry about.

### *Spots*

Discoloration and spots are very common symptoms. When we see spots it is important to look at them closely to find out what color they are as well as whether they are raised, smooth or fuzzy.

## WHITE SPOTS

If the fish is coated in small white spots that makes it look like the fish has been sprinkled with sugar or salt, the fish has a parasite called ich. However if the spots are larger and more blotchy, they are probably flex (aka columnaris).

### Causes and Treatment

### *Ich*

White spots that look like sugar coating, indicate that the fish has ich, like in this picture. Ich is the common name for the ciliated parasite Ichthyophthirius mutifillis, which has a complicated life cycle. If left untreated, ich is deadly. Treat the whole tank with Rid Ich (following the instructions on the bottle) or Quick Cure. If it is a goldfish, you can treat with

### *Flex or Columnaris*

White fuzz on fins and body is most likely a bacterial illness, flex/columnaris caused by a bacteria Flavobacterium columnare (previously called Flexibacter columnare). Flex/columnaris can be a very quick killer. Fuzz is not always how flex manifests itself. Sometimes there may be redness and deteriorations or sores. Redness in the mouth area is sometimes flex. Flex may also be seen as greyness or yellowness to the slime coat of fish.

It is important to isolate the fish in a hospital tank to eliminate spreading. If the fish is salt tolerant, add 1 tsp of salt per gallon of water in the hospital tank and to your main tank as well (to help prevent infection and spreading). Flex is an opportunistic infection, and mild cases will clear up with super clean water and a little salt. In advanced cases, the fish will need antibiotics.

Since flex is mostly an external infection, bath antibiotics may be effective such as Maracyn Plus (sulfadimidine and trimethoprim), Jungle Fungus Clear (nitrofurazone, furazolidone, potassium dichromate), and kanamycin. Potassium permanganate is a very effective flex treatment, but it must be used with caution as it is a strong oxidizer and could put a very sick fish "over the edge." Some strains of flex can be hard to treat.

### *Breeding Stars*

Male goldfish get breeding stars on their operculum and first ray of their pectoral fin. These are natural, so no need to worry or treat.

### *Mystery Spots on Goldfish Fins*

On the Goldfish and Aquarium Board forums we have seen a few goldfish with spots or small pimple-like bumps on their tails that may resemble columnaris but which do not respond to treatment with either parasite medicatons or antibiotics. They do not appear to cause the fish discomfort and seem to come and go. At this point most of us think they may be caused by a viral agent but we do not really know.

## YELLOW SPOTS OR SHEEN

If your fish looks like it is getting a golden brownish sheen to it, you are dealing with another parasite.

### Causes and Treatment

#### *Velvet*

If there is a coating of yellow (or brown) on the fish, the fish has velvet. This is a parasite - a dinoflagellate called Piscinoodinium (formerly Oodinium). The fish may also have clamped fins or scratch against tank decorations or gravel (flashing). Once the attack is severe, the fish may become lethargic and you may see rapid gill movements. This parasite will attach to and attack the skin or gills of fish. It is deadly and quite contagious, so treat the whole tank with Rid Ich or Quick Cure. Sometimes velvet can be difficult to diagnose, and it may help to shine a flashlight on the fish to see the yellowish coating.

## BIGGER RAISED DARK SPOT

If you see green or brown bumps on your fish, and the fish is acting like it has parasites (flashing, clamped fins), study the bump carefully to see if it has black dots on it or if it moves.

### Causes and Treatment

#### *Argulus*

If you see the eye spots, your fish has a parasite called Argulus , also known as fish lice.

Argulus is a disc-shaped crustacean parasite (a copepod) that gets up to 10 mm in diameter. It will attach itself to the fish with suckers and suck blood from the fish. This is very irritating, and the fish will normally flash.

Treat by removing the lice with tweezers if possible. If the removal leaves a sore, you can dab quickly with some hydrogen peroxide or iodine/betadine to prevent infection. Lice can live several days (maybe as long as 15) without a host and are egg-layers, so to make sure all lice have been removed from the tank environment, you may want to treat your tank with dimilin (found in a medication like Anchors Away). Another effective treatment is the chemical trichlorfon, found in products like Clout and Fluke Tabs. Trichlorfon is a neurotoxic pesticide and must be used with caution.

## RED SPOTS

Red spots anywhere on the body of your fish.

### Causes and Treatment

#### *Costia*

Small red spots (especially on the underside of the fish and companioned with flashing and rubbing) could be costia - the protozoan parasite Ichthyobodo. Sometimes costia will not always show up like this, but rather presents as slime coat changes and perhaps some blueness or blackness to the fish. A clear diagnosis sometimes requires a microscope. Costia should be treated with Rid Ich or Quick Cure. Though it has a short life cycle, it can be quite hard to treat and treatment requires several applications.

#### *Internal Infection*

If the redness is in the form of blotches or red streaks, the fish may have an internal bacterial infection, usually caused by gram negative bacteria such as

Aeromonas or Pseudomonas. Often the redness can be found at the base of the fins. Treat by feeding antibiotic food (Medi-Gold preferably) or if the fish is not eating, use a full spectrum water antibiotic like kanamycin, acombination of Maracyn (erythromycin) and Maracyn 2 (minocycline), or Maracyn Plus (sulfadimidine and trimethoprim).

#### *Other possibility*

If the fish is not showing other signs of parasites, red spots may be caused by slight irritation, rubbing, or scratching and the best thing to do would be to check the water quality and just watch the spots. If the spots are open, they would be considered sores

## BLACK SPOTS

Black on the fish is often less of a concern than the other spots we see.

### Causes and Treatment

#### *Melanophore Migration (or healing)*

Black is normally an indication that the fish has suffered a chemical burn, and this burn is now healing. This is sometimes called "melanophore migration." If you see black, check your water and perform water changes to correct problems.

Goldfish with black spots caused by poor water quality. These spots disappeared when the fish recovered in water with perfect quality.

#### *Uvulifer amblopitis*

Much rarer are the black spots caused by a parasite (a flatworm called Uvulifer amblopitis ). Since this parasite's lifecycle requires intermediate hosts (like birds), it is found only in ponds.

## SPOTS OR CRATERS ON HEAD OR ALONG LATERAL LINE

Here is a picture of some of these craters.

### Causes and Treatment

#### *Hole in the Head Disease*

These kinds of spots are associated with a disease called Hole in the Head Disease (HITH) or HLLE (Head and Lateral Line Erosion). It is a disease that affects cichlids in particular but is still not clearly understood. Several causes have been explored such as water quality, mineral or vitamin deficiency, infection or parasite infestation. (Hexamita). Treatment is therefore difficult. But we would recommend checking water, addressing food issues, feed Metro-Med (to deal with Hexamita).

## WORM ATTACHED TO FISH

Of the common parasites only anchor worms, leeches, and fish lice are big enough to be easily seen with the naked eye.

### Causes and Treatment

#### *Anchor Worm*

If you see a worm attached to the fish with forked ends, it is an anchor worm, a parasite also known as Lernea.Treat by removing the anchor worm with tweezers if possible. If the removal leaves a sore, you can dab it quickly with 3% hydrogen peroxide or iodine/betadine to prevent infection. Anchor worms are egg layers, so to make sure your tank is free of worms, you may want to treat your tank with dimilin found in a medication like Anchors Away.

#### *Leeches*

Leeches or Hirundea are flattened, segmented worms with a sucker on both tail and mouth. They move by attaching the mouth sucker, then tail sucker, then mouth sucker, and so on, to the surface they are on. They can latch on to the fish and cause harm and irritation. They are not easy to remove. A salt dip is a possibility and there are some commercial remedies available. Leeches are not common in aquariums but can attack fish in ponds.

## RED WORMS PROTRUDING FROM THE VENT

Fish can suffer from several kinds of internal parasites. This is a condition that is hard to diagnose. Often failure to thrive is the only clear indication we get. However, one common internal parasite is different in that we can observe it protruding from the vent area of the fish as shown in the picture below.

### Causes and Treatment

#### *Camallanus*

If you can see a red worm protruding from the anus (vent), your fish has an internal parasite called Camallanus. Internal parasites can come from live food or be transmitted from fish to fish. This roundworm (nematode) is easy to identify because of its tendency to hang out the vent. Nematodes are best treated with fenbendazole, piperazine, or levimasole. Jungle Anti-Parasite Medicated Fish Food contains levimasole.

## FUZZ, COTTON GROWTH

It is not unusual to see fuzz growing on fish, and when we do, it is clear that the fish in most cases is ill. This fuzz is sometimes described as cotton wool and some people refers to fuzz as cotton wool disease. Others call all fuzz fungus. But only rarely is the fuzz or cotton growth that we can observe caused by fungi; in most cases it is caused by a bacterium.

## Causes and Treatment

### *Columnaris*

White fuzz on fins and body is most likely a bacterial illness, often called flex, columnaris, or cotton wool disease that is caused by a bacteria--Flavobacterium columnare (formerly called Flexibacter columnare). Columnaris can be a very quick killer, especially when it infects the gills. Flex often looks fuzzy--like cotton wool--but not always. Sometimes there may be redness and ulcers/sores. Redness in the mouth area is sometimes flex. Flex may also be seen as greyness or yellowness to the slime coat of fish. Another symptom is a saddle back lesion where the fish turns gray in the upper part of its body.

It is important to isolate the fish in a hospital tank to reduce the risk of the disease spreading to other fish. Some strains of columnaris can be hard to treat. If the fish is salt tolerant, add 1 tsp of salt per gallon of water. Adding salt to your main tank will help prevent infection because it keeps flex from being able to stick to the fish. The fish will need antibiotics. Several antibiotics may be effective against columnaris - erythromycin and oxytetracycline are traditionally thought to have really good effects. Other options are sulfa drugs, nitrofuran antibiotics, and potassium permanganate.

### *Treatment Options*

If the fish is still eating, choose one of the following options:

1. Feed Jungle Antibiotic Food (can be combined with Jungle Fungus Clear bath treatment) or
2. Feed Medi-Gold or Romet B (can be combined with kanamycin bath) or
3. Feed Metro Med

If the fish is not eating or eating poorly, choose one of the following options:

1. Add Maracyn I combined with Maracyn II to the water or
2. Add Jungle Fungus clear to the water or
3. Add kanamycin or kanamycin sulfate to the water

If you are an advanced fish-keeper, you may consider using potassium permanganate (PP) at the start of a treatment regimen to help knock down the bacterial load on the fish. Please consult with us on the Help! My fish is Sick! Forum or published texts for guidence on using potassium permangante. Because PP is an oxidizer, it can cause gill damage.

### *True Fungus*

If the fish has been injured and you see fast growing white/gray fuzz, this is most likely fungus and should be treated with Maroxy (treatment for true fungus).

If it is a large fish, you can skip the whole tank treatment and just dab the affected areas with iodine/betadine or malachite green.

### *Fungal Spots on Wen*

Goldfish with wens may get fungal spots on their wens. Some times these spots are called 'growth spots'. Usually intervention is not needed, but occasionally if it looks bad you can clean it with some 3% hydrogen peroxide or some iodine/betadine on a Q-tip (taking care not to get peroxide or iodine in the gills or eyes). Depending on how cooperative your goldfish is, you may need to sedate it before you do this.

## SORES

Red sores are best diagnosed by knowing the fish's history. If the fish has suffered trauma the red sores are actual wounds, but if the sores slowly appear they are likely caused by bacteria and are developing ulcers. Redness in the mouth area is probably flex

### Causes and Treatment

#### *Ulcers*

Red sores on the body are often signs of a bacterial infection and ulcers. The bacteria that causes these sores is often Aeromonas hydrophila. This type of infection is often secondary to fluke infestation, so we recommend treating for flukes as well as addressing the sores directly. It is important to isolate the fish in a hospital tank.

You can treat topically as explained and combine this treatment with feeding of medicated food. Medi-Gold and MediKoi are the best choices for treating this kind of infection. If there is a white center developing, it is clearly infected and may be ulcerating. The fish may need to be injected with an antibiotic, or if that is not possible treated in a bath of Tricide Neo (sometimes sold as Neocide 3).This is a serious illness, but treatable, so please post in our forum so we can help you more specifically.

#### *Cuts and Scrapes*

If the sore is the result of an injury, the best thing to do is to keep the water clean and feed good food so that the fish will nicely recover on his own. The sore can be swabbed one time with Iodine or 3% hydrogen peroxide. Additional swabs will kill the new, healthy cells growing in. If the fish has suffered some trauma and is bleeding, pressure can be applied to the wound to stop the bleeding.

## RED STRIPES OR STREAKS

Red streaks/stripes/veins in the fins, often the tail fin.

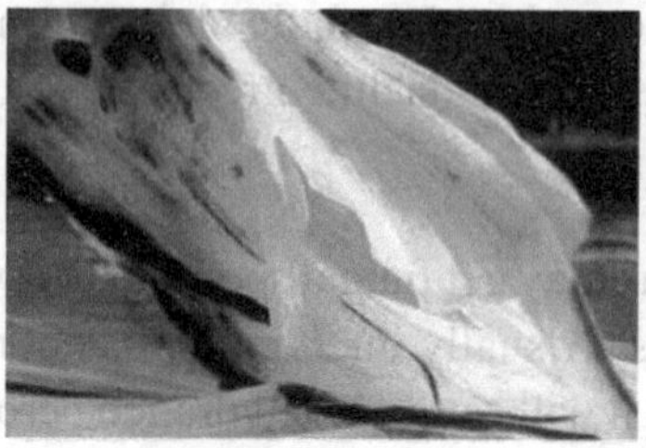

## Causes and Treatment

### *Water Quality*

Streak are often caused by poor water quality. Check your water parameters and perform water changes if necessary. These tests are best done with your own testers. As a minimum you need to test for ammonia, nitrite, nitrate, and pH. If you need help interpreting the tests, please visit GAB's Water Quality Forum. Some people associate red streaks with either high nitrate or with low oxygen. In either case a partial water change should help.

### *Natural*

On some fish (goldfish in particular) that are very light in colour, you can see bloodlines on their tails. This seems to be just a natural thing so if keeping nitrates low (under 40) and having plenty of aeration do not help, and the fish is acting healthy otherwise, consider these streaks normal.

### *Bacterial Infection*

If attending to water is not helping and the fish is looking lethargic, it is likely a bacterial infection. Treat by feeding antibiotic food (Medi-gold preferably) or if the fish is not eating, use a full spectrum water antibiotic like kanamycin or a combination of Maracyn and Maracyn 2. If the fish is larger, you should consider injections.

## FIN ROT OR DETERIORATION

Fins are vulnerable to poor water quality, aggression and to bacterial problems. If you see deterioration of fins, you need to analyze the situation carefully. Many people over-treat fin problems with antibiotics, but checking and adjusting water quality or looking for aggression between the fish is the first place to start. Some fish, like bettas, may blow their fins and there is no reason to medicate but plenty of reason to see whether decorations, tank mates or environmental stress factors are the main cause of the problem.

Picture is showing redness (1),red area has sloughed off (2) and healing (3) Finrot on Betta. Only the red areas are finrot

***Treatment and Causes***

***Water Quality***

Fin rot is often caused by poor water quality Check your water parameters and perform water changes if necessary. These tests are best done with your own testers. As a minimum you need to test for ammonia, nitrite, nitrate, and pH. If you need help interpreting the tests.

***Stress or incompatible tank mates***

Some fish (like Bettas) react to stress by blowing their fins (or even sometimes nip at their own fins). Stress can be caused by sudden changes in pH or water temperature. Many fish are fin nippers (like Barbs, Serpae Tetras) and will cause fin problems. However in most cases clean water and good conditions will keep up with the nipping.

***Bacterial Problem***

If the ends are red and/or you are seeing very rapid fin loss, you may have to treat for bacteria (Aeromonas, Pseudomonas, Flavobacterium) . If the fish is bigger and you can handle it, swabbing the fins with peroxide or iodine and applying triple antibiotic cream may be best--combined with medicated food. If the fish is small, feed medicated food or use water antibiotics if it is not eating. If the fish is larger, you should consider injections.

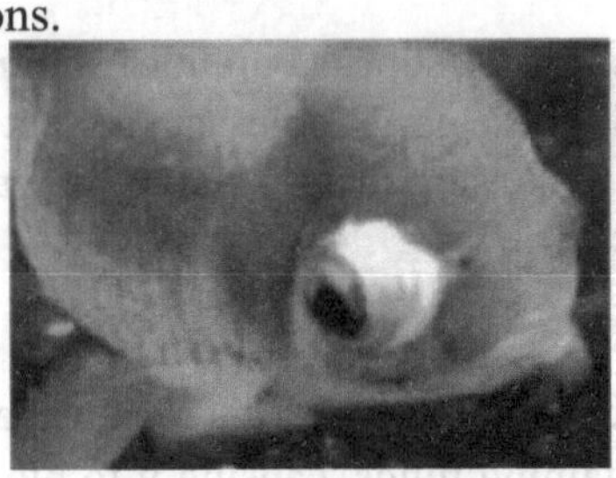

## EYES BULGING OUT/POPEYE

Popeye means that one or both eyes bulge out abnormally. This is caused by accumulation of fluid either in the eye itself or behind the eye. Pop eye is a very complicated condition, mainly because it is a symptom of an underlying problem. Part of the challenge of treating is to try to find the underplaying cause which can be poor water quality, bacterial problems, nitrogen supersaturation (gas bubble disease) and/or nutritional deficiencies (according to Edward Noga).

### Causes and Treatments

***Water Quality***

Popeye may be caused by poor water quality. Check your water parameters and perform water changes if necessary. These tests are best done with your own testers. As a minimum you need to test for ammonia, nitrite, nitrate, and pH. If you need help interpreting the tests

### *Bacterial Problems*

In addition to water quality, an internal bacterial infection is often the cause. However it can be hard to clearly diagnose such an infection. The bacteria that could cause popeye may be Aeromonas, Pseudomonas, or Edwardsiella, all of which are gram-negative rod bacteria. In some cases these bacteria may also cause red blotches or red streaks on the body and fins. You may be able to spot an internal infection early if the area around the pectoral fins is red. It is when the bacterial infection attacks the internal organs that the eyes begin to collect fluid. If your fish is still eating, treat by feeding antibiotic food. The most effective medicated food is Medi-Gold, which contains three antibiotics, including kanamycin. If the fish is larger, you should consider injections

### *Diet*

Experts have identified nutritional deficiencies as a cause for popeye. These deficiencies may also be seen as failure to thrive and other signs of starvation. It is important to pay attention to the nutritional needs of the fish and to feed a varied diet.

### *Gas Bubble Disease*

In rare cases, excess levels (super saturation) of nitrogen in the water may cause gas bubbles in the blood. These gas bubbles can collect behind the eyes causing the popeye. To prevent this make sure there is plenty of water agitation as you add the water and use air stones.

### *Dropsy*

Sometimes popeye is the first sign of a serious bacterial illness that will lead to dropsy. The fish is retaining fluids, causing it to bloat and get raised scales that make the fish look like a pine cone. It is very hard to save a fish once this happens. Most cases are fatal. Goldfish in a few cases have responded well to Metro-Med, so if the fish is eating this is worth trying. Other have tried antibiotic injections to try fight bacteria.

### *Other Causes*

Experts (such as Edward Noga) have identified parasites and viruses as possible causes of popeye. Certainly none of the common parasites in fish tanks will cause popeye so these are much rarer cases.

Doc Johnson from Koivet has identified Fish TB as one possible cause of popeye .

There are times when injuries and trauma may have the effect of making the eyes pop out or in some cases a tumor growing behind the eyes.

## CLOUDY EYES

Eyes can be described as cloudy if the outer covering is milky-colored or opaque or if the fluid inside the eyeball is cloudy. Sometimes there may also be a cloudy fungus film on the outside of the eye.

### Causes and Treatments

#### *Water Quality*

If your fish develop cloudy eyes, the most likely reason is poor water quality. Check your water parameters and perform water changes if necessary. These tests are best done with your own testers. As a minimum you need to test for ammonia, nitrite, nitrate, and pH

#### *Bacterial Illness*

It is possible that bacteria may cause cloudy eyes. Look for other signs such as loss of appetite, bottom sitting, redness at the base of the fins, or popeye. If the fish is not eating, use water antibiotics. If the fish is larger, you should consider injections

#### *Injury*

Quite often cloudy eyes are caused by trauma such as netting, handling, or injury on sharp objects in the tank. Make sure you are as gentle as possible if you must handle fish and take steps to remove sharp objects that the fish may injure themselves on.

#### *Old Age*

Some fish may turn blind with old age. Good nutrition and healthy tank conditions may help the fish keep its sight well into old age.

#### *Eye flukes*

These are not very common in aquariums. The most common eyeflukes (Diplostomum spp.) require birds as part of their life cycle.

## SLIMECOAT CHANGES

Fish have a slimecoat (a mucoprotein protective layer) that covers their scales . This coat provides protection against cuts, scrapes, and invasive pathogens such as parasites and bacteria. The enzymes and antibodies in the slimecoat helps kill invading organisms, so the slimecoat is truly the armour of our fish.

Because of the slimecoat it is important to handle the fish as little as possible, and when we have to handle them, be as careful as possible. At GAB we do not believe in adding any of the products that claim to help strengthen slimecoat. There is little evidence that such products do much good. We worry about what 'unnatural aquarium products' like aloe vera may do to the gills of fish. An exception to this may be products that contain vitamin B12, which may help to strengthen the fishes own ability to produce slimecoat.

Fish can develop excess slimecoat. This is an immune response and usually occurs because something is irritating their skin. Excess slime can either be seen with the naked eye (sometimes even in the form of stringy mucus hanging off the fish), or you can check for excess slimecoat by touching or by running a popsicle stick (wooden or plastic) gently along the side of the fish.

## Causes and Treatments

### *Water Quality*

Excess slimecoat is very often caused by poor water quality. Poor water may cause the fish to strengthen its defense system. Check your water parameters and perform water changes if necessary. These tests are best done with your own testers. As a minimum you need to test for ammonia, nitrite, nitrate, and pH. If you need help interpreting the tests.

A change in pH causes stress and often changes to slimecoat, particularly in goldfish. If you see grey slimecoat, pH is the first thing you ought to check.

### *Salt levels*

For most fresh water fish, high levels of salinity in the water can cause irritation and lead to excess slimecoat production. Some people believe that by adding salt and causing the fish to produce this excessive slime coat they are helping the fish ward off pathogens and stay healthy. However the effect is likely counter productive as the fish is stressed when it produces more protective slime, and stress negatively affects both osmoregulation and immune response. Some fish (like goldfish) can tolerate low levels of salt (up to 0.3%) very well and salt can be used to cure some parasites like ich.

### *Parasites*

One of the major reasons fish produce excess slimecoat is in response to external parasites such as ich, costia, and flukes.

If you can not see other signs of parasites (see symptoms on spots and flashing) and the water is fine but the fish is producing excess slimecoat, you may want to treat with Quick Cure or Rid Ich, as the cause is likely the parasite costia.

### *Other irritants*

It is possible for excess slimecoat to be caused by household irritants such as sprays and cleaners that may enter the water. Make sure that you do not use any cleaners near the aquarium and that your hands are clean if you stick them in the water. Adding fresh carbon will help remove chemicals that may have entered the water.

Sometimes fish may be irritated because they are rubbing against ornaments or decorations in the tank constantly. So make sure you plan for plenty of swimming room for your fish.

### *Gills: Paleness or other problems*

Gills are vital organs as they are responsible for respiration, excretion of wastes, and they have a role in maintaining fluid and electrolyte balance. Gills are made up of threadlike structures called filaments. Each gill filament has secondary lamellae protruding off of it like a christmas tree with branches. The epithelium covering these structures is only one or two cells thick. The gill filaments contain a network of capillaries that result in a large surface area in contact with the water.

Because they are so vascular, gills are susceptible to damage from water quality and are targets for both bacteria and parasites that can cause severe problems for the fish. Symptoms associated with gill damage include oxygen seeking behavior (e.g. rapid gill movements, flared gill covers, piping or gasping at the water surface), excess mucus, lethargy, lack of appetite, or irritation (e.g. rapidly opening the mouth and fluffing the gills).

When fish appear to be sick, looking at the their gills may provide clues as to what's going on. Healthy gills should be blood red (or cherry red) in most fish. If the fish is big enough, you can gently open the gillcover and inspect the gills.

### *Picture of Healthy Gills*

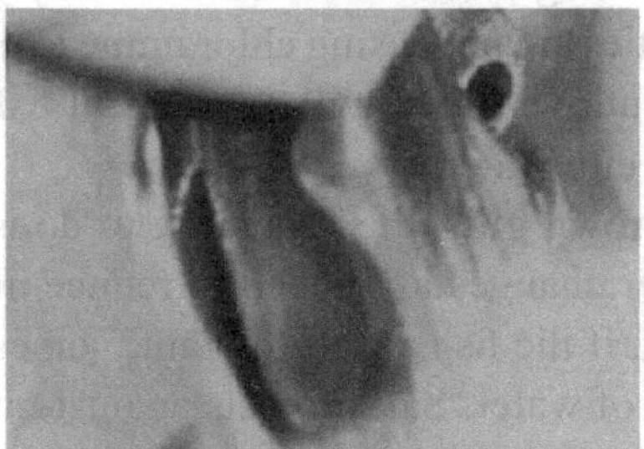

Healthy gills in most fish should have a bright blood-red colour and the filaments should be uniform and well defined. When you look at them, you should be able to see the filaments clearly. Signs of problems include: brown or pale gills, blotches or stripes, or the gill filaments are not clearly divided (clumped together) or are worn or torn at the ends.

***Picture of Gills***

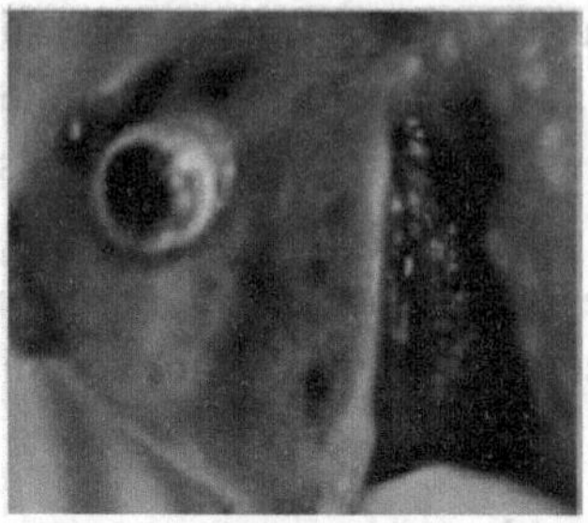

This picture gives a good indication of how to look at gills in a bigger goldfish. The goldfish is held over water and partially in the water. The picture also shows off the nice dark red colour that goldfish gills should have when healthy.

## DIFFERENT GILL ABNORMALITIES AND POSSIBLE CAUSES

### Pale Gills

If the gills are pale, the fish may be experiencing anemia or organ failure. If the gills are pale and the fish has red streaks on its body or fins, is bloated or dropsying, organ failure is very likely. Paleness may also be caused by blood loss caused by parasites like flukes attacking the gills. If possible, look at a mucus scrap under the microscope and attempt to determine whether parasites are involved. Flukes are common in goldfish, and can be treated with praziquantel. For other external parasites a malachite green/formalin combination is usually effective.

### Brown Gills or Brown Blood Disease

Converted to methemoglobin which can not carry oxygen, so the fish can suffocate. Most larger cities are now using chloramine to disinfect drinking water, so letting the water sit out over night will not make it fish safe. Be sure to use a dechlorinator if you use city water.

If your fish have brown gills, the first thing to do is to test your water for nitrite and perform water changes as needed to reduce nitrite levels. To protect against nitrite poisoning, if the fish is salt tolerant, you can add salt to the water at 1 teaspoon per gallon of water. Salt protects against nitrite poisoning in some fish (with chloride cells in their gills), because the chloride competes with nitrite for uptake via the gills. Methylene blue can be used in cases of nitrite poisoning to unbind the red blood cells. If you are seeing nitrite, you should re-evaluate your stocking and filtration as nitrite should not be present in established tanks.

### Filaments 'stuck' together or Hyperplasia

Hyperplasia (excessive cell proliferation) occurs in response to irritation. This excess cell growth is typically combined with excess mucus production causing the lamellae to clump together, thereby reducing gas exchange. The clumping and excess mucus may also protect parasites from the effects of medicines. In cases of hyperplasia, is important to test the water, do partial water changes if indicated, and add fresh carbon. Irritation may also be caused by parasites, you could treat with praziquantel for flukes and/or malachite green/formalin for other external parasites.

Once the lamellae clump together this will affect gaseous exchange and respiration negatively.

Pictures (microscopic) of Gill filaments stuck together

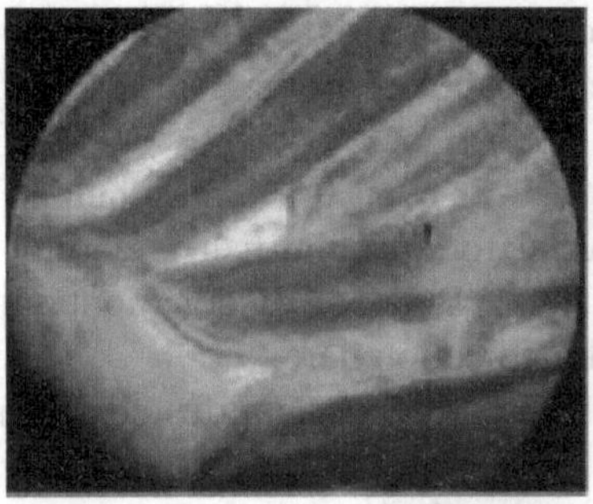

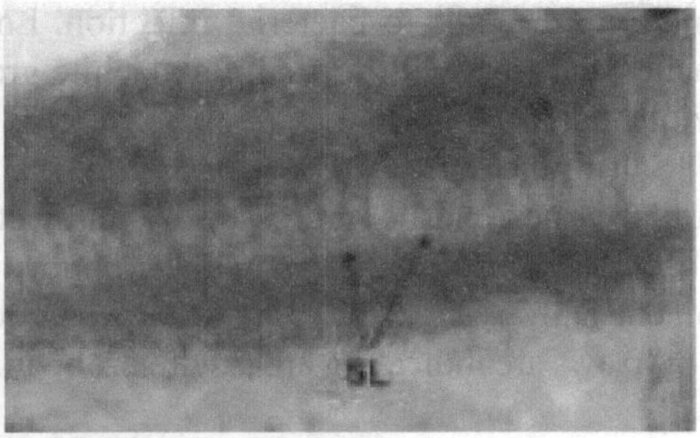


Here the secondary lamellae (SL), the parts of the gills that exchange oxygen and carbon dioxide, are clumped together by mucus.

## Frayed Gills

If the gills are frayed in the ends, they have probably been 'burned' by chemicals like ammonia or by very low pH. Sometimes formalin or potassium permanganate can cause fraying. If you see fraying and no other symptoms, it is probably best to keep the water pristine and let the gills heal on their own.

## Bleeding from Gills

If the fish has been recently netted or handled, this could cause bleeding. Another likely cause is water quality or parasites (like flukes), but it could also be bacterial gill disease. Make sure the water is pristine, run the fish through a parasite treatment (flukes and other parasites), and feed medicated food (medi-gold preferably).

Gills clumped together and eroding. Yellow color on gills or hamburger-like gill consistency

If the gills are clumped together and eroded, often with yellow patches, this is likely bacterial gill disease. Typically this is caused by bacteria (e.g. Flavobacterium columnaris) after the gills have been damaged by water quality issues or parasites.

Pictures (microscopic) of gills from a Celebes rainbowfish with bacterial infection and possible ammonia damage

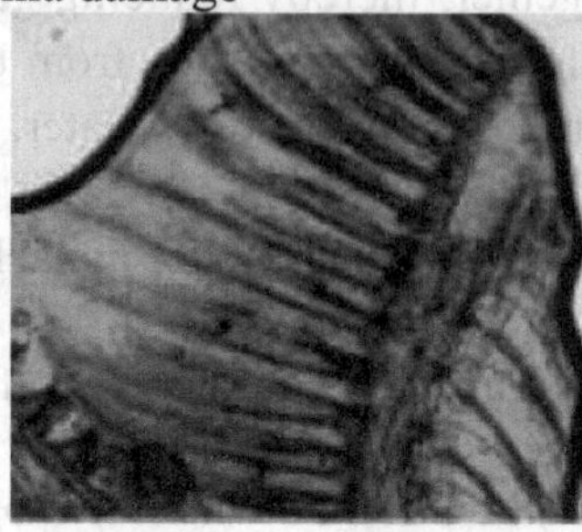

Treatment should include treating any parasites that caused tissue damage opening up the gills to bacterial infection. Look for other external symptoms of parasites and treat with formalin/malachite green. If flukes are present, treat with praziquantel.

Salt in the water (0.3%) for salt tolerant fish or even 0.1% may help. The bacterial issues can be addressed by antibacterial food, if the fish is eating, or antibacterial water medication (like Maracyn Plus). BGDX Powder (containing p-toluene sulfonyichioride) may also be a good medication here.

Arrows are indicating areas with problem. The problem is manifested as black mass.

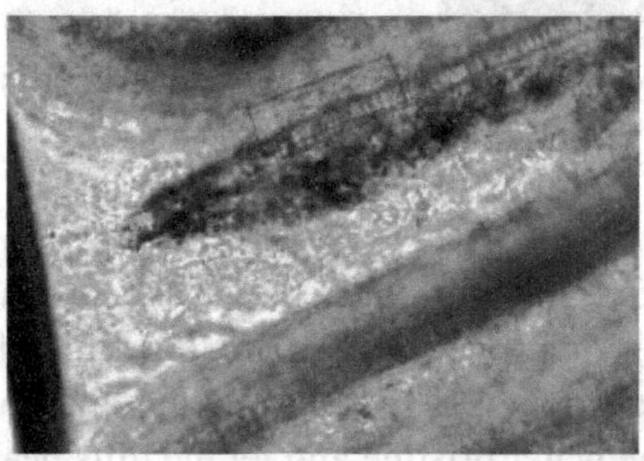

Close-up of same gill. The blue box is around columnaris "haystacks"

## STRINGY WHITE POOP

Diagnosing fecal problems is not always easy. But for most fish, white stringy poop is a sign of a problem and something to take notice off . However, poop will be the color of the food fed so if you are feeding white, it is probably natural.

### Causes and Treatment

#### *Stress*

Stress can cause white poop. For a fish, stress is bad water quality or changes in water parameters. Measure for ammonia, nitrite, nitrate, and pH. Check the temperature. Aggressive tank mates may also lead to stress, as can illness and medication. Moving to a new tank or the addition of new tankmates can also cause stress.

***Hexamita***

The parasite hexamita, that can attack internal organs, very often causes white stringy poop. Feed the fish anti-parasite food that contains metrodonidazole.

***Infection***

Bacterial infections can lead to stringy white poop. Before acting check for other signs of infection (redness, lethargy, lack of appetite).

# FLOATING AND SINKING

There are times when we can find our fish either sitting on the bottom incapable or hesitant to swim around, or other times the fish is floating at the top sometimes even flipped over. These problems of buoyancy are complex and often quite serious.

## Causes and Treatments

***Water Quality***

Floating or sinking can be caused by poor water quality. Many people think that high nitrate may cause floatiness in goldfish. So if you notice buoyancy problems, check your water parameters and perform water changes if necessary. These tests are best done with your own testers. As a minimum you need to test for ammonia, nitrite, nitrate, and pH. If you need help interpreting the tests.

***Constipation***

If your fish lose control of their swimming, they may be constipated. Some fish that are constipated may bottom sit. Take care to check the nutritional requirements for your fish. Do not over feed. If you think constipation could explain the symptom, fast your fish for 3 days. You may try to feed a piece of pea after the fasting period.

***Gassiness***

Many people believe that when fish get floaty, excess gas in their digestive system may be the cause. Fish could get gassy by swallowing too much air or by eating food that releases gasses in the digestive system. If the fish is floaty, avoid feeding foods that may cause gas (such as soy or wheat products). Floaty poop may be a further indication of gas in the GI tract.

***Infection***

Illness can cause buoyancy problems. Internal infection can interfere with the fish's ability to regulate gas density in the air bladder and cause buoyancy problems. If there are signs of infection (e.g. lack of appetite, lethargy, redness, etc), feed medicated food, or if the fish is not eating add a broad spectrum antibiotic to the water. For larger fish, antibiotic injections are preferred. Bacterial infections are often secondary to wounds caused by parasites such as flukes, so if you have access to a microscope, doing a scrape and scope to look for parasites is recommended.

***Air-Bladder malfunction***

The air-bladder (or swim bladder) is not a very well understood organ. It seems to be linked to sensory functions like hearing, and some fish use it to produce sound. For fish that have air bladders, the organ is also used to control buoyancy. The gasses inside the bladder (which are not air) are compressed or decompressed (probably through the tissue outside).

It is possible that ailments (bacteria, parasites, virus) may cause buoyancy problems, either by affecting the small blood vessels that feed gasses into the swim bladder or because as the fish is too weak to use the muscles around the air bladder. Water quality (like high nitrite or nitrate) may also cause the fish to experience buoyancy problems. Internal cysts and tumors may also cause swim bladder malfunction by physically compressing the organ.

***Genetic problems***

Some fish (like fancy goldfish) are bred to have short round bodies and this may cause the fish to have less room for vital organs such as the air bladder. As a result the fish may experience buoyancy problems. Some fish may also have deformed swim bladders.

## LOSING WEIGHT

If your fish fails to thrive or is losing weight despite eating, you are looking at a couple of likely explanations.

### Causes and Treatments

***Internal Parasites***

If you are feeding regularly and the fish is still losing weight, you may be dealing with internal parasites such as nematodes, cestodes, or flagellates. Treat with Jungle Antiparasite Food which contains metronidazole, praziquantel, and levamisole. Praziquantel and levamisole are also effective de-wormers as bath treatments.

***Hexamita***

Hexamita (a flagellate, Spironucleus vortens) is a common internal parasite in some species but hard to diagnose. Weight loss, gill fluffing, and spitting out food are common signs. The best treatment for hexamita is metronidazole available in products such as Jungle Antiparasite Food or Metro-Med.

***Fish Tuberculosis***

If treating for internal parasites does not help, there may be more serious issues like fish TB (more correctly called mycobacteriosis) or other viruses. If that is the case, the fish should be isolated. Mycobacteriosis is incurable.

## LOSING COLOR

If the colours of your fish are less vibrant or if they turn pale, this may be a sign of problems.

### Causes and Treatments

#### *Water Quality*

Loss of colour may be due to poor water quality. Check your water parameters and perform water changes if necessary. These tests are best done with your own testers. As a minimum you need to test for ammonia, nitrite, nitrate, and pH. If you need help interpreting the tests.

#### *Stress*

Stress is often a factor in fish changing colour. Stress can occur in response to stressors. Common stressors include: handling, changing water parameters, water changes, aggressive tankmates, changes in temperature or the wrong temperature, noise, vibration, sudden movements, or overstocking/crowding. Try to identify the source of the stress and alleviate it. Stress often leads to illness because it depresses immune response.

#### *Bacterial Illness*

Some fish get pale when fighting an illness or an infection. Make sure you look for other signs of illness (e.g. redness, lethargy, loss of appetite, etc) and medicate appropriately with medicated food if the fish is eating or broad spectrum bath antibiotics if the fish is not still eating.

#### *Costia*

Costia is a protozoan parasite, Ichthyobodo, that attacks the fish's skin. This attack often causes the fish to produce more slimecoat, and as a result the fish may look whiter or paler. Read here about Bugsy the black moor's costia to see good pictures of scopes of the parasite. If the whiteness is caused by costia, you should be able with closer inspection to see the slimecoat changes.

The fish will often also flash and rub because of the irritation. Costia is best treated with a malachite green/formalin combination present in parasite medications like Rid Ich or Quick Cure.

#### *Natural Colour Changes*

Some fish, like goldfish, will change colour over time. Goldfish may lose their orange, red, or black colour due to environmental changes or genetic makeup.

### *Neon Tetra Disease*

Tetras and perhaps other cyprinids are susceptible to a sporozoan parasite named Pleistophora hyphessobryconis. The parasite attacks the intestines and eventually muscle tissue of the fish and will cause paleness. Neon tetras, for whom this disease is named, will lose the blue colour in their middle colour band.

There is no known treatment and it is recommended to euthanize infected fish to prevent spreading. However neon tetras often react to stress by paling, so be sure you have eliminated other causes before taking this step.

# BENT SPINE

If you look at your fish from above and you see a bent spine or an S curvature, there are several possible causes.

## Causes and Treatments

### *Birth Defect*

Bent spine is a common genetic birth defect and should be culled out. Fish with this condition should never be allowed to breed. Sometimes instead forming an "S" shape, birth defects result in a "U" shaped spine.

### *Malnutrition or Scolosis*

Malnutrition may cause spinal deformities so take care to check your fish's nutritional requirements and feed fish every day. Lack of Vitamin C is sometimes identified as a cause of scolosis, so take care to properly store your fish food and be sure it is not too old. A good rule of thumb is to toss out any foods that have been opened for 6 months. Fresh food may be a good source of vitamin C as well.

### *Injury*

Fish may become injured from fleeing (fight or flight reactions), jumping against things when spawning, aggression, during tank maintenance, etc. If this occurs, keep the water pristine and reduce the cause of the aggression or flight response. In severe cases fish can break their spine.

### *Electricity*

If your fish is outside in a pond, a bent spine can be caused by lightening strikes. Equipment leaking electricity in the tank can also result in a bent spine if the fish becomes grounded.

### *Myco-bacteriosis or Fish Tuberculosis*

A bent spine is one of the signs that the fish in infected with bacteria called Mycobacteria. Some people refer to this condition as "fish tuberculosis" or "fish TB." Mycobacteria (there are several kinds) are gram-positive bacteria that infect slowly and gradually cause more and more problems for the fish. Other symptoms may include emaciation, redness, lethargy and exophthalmia (pop-eye). If this infection spreads to the skeleton of the fish, the spine may bend. There is not currently any treatment for mycobacteriosis. Isolation or euthanasia are recommended to avoid spreading.

Many people believe that fish that are over crowded and lack nutrious food are far more susceptible to this infection than fish kept in good conditions. So it is a disease easier to prevent than cure.

Mycobacteria can cause "granulomatous skin disease" or "fish handlers disease" in humans. It enters through breaks in the skin. If you suspect fish TB, avoid placing hands or arms with open sores in the aquarium. Cleaning hands with anti-bacterial gel hand sanitizers should be done as a precaution. People with a compromised immune system should avoid coming in contact with aquariums where fish TB is suspected.

***Other causes***

Internal swellings, tumors, or not being able to expel eggs can also cause bent fish because the pressure or pain from the swelling/tumor causes them to favor one side.

## BLOATING

If fish get bloated (swollen in the belly area), you may be dealing with constipation, gas, or retention of fluids.

### Causes and Treatments

***Constipation***

Constipation may happen if the fish is fed too much or fed the wrong foods. Fasting may help here. Review the food you are feeding and make sure you're feeding the correct type of food for your fish (e.g. omnivore, herbivore, carnivore). Eliminate foods that may cause gassiness (soy may cause gas in the gastrointestinal tract). Try adding more fresh foods to the diet. Feeding a pea sometimes helps. If peas don't help, you can add a small grain of Epsom salt inside the pea.

***Fluids***

Check the gills of the fish. If they look pale, organ failure is likely. Organ failure can cause the accumulation of fluids in the viseral cavity (where the organs are). Also look for signs of onset of dropsy.

***Eggs or egg bound***

Female fish that carry eggs may also get very round. In rarer cases if the eggs are not released or expelled, the fish may get "egg bound" or egg impacted. There may not be a cure for this. With larger fish (like large goldfish) one may be able to stimulate egg laying. If a fish is truely impacted, she will need surgery. There is some anecdotal evidence that feeding fish a grain of epsom salt may help.

***Kidney problems or internal dysfunctions***

Bloating can be caused by serious internal problems. There may be an infection or a tumor, or the kidneys may not be working properly. There cases are hard to diagnose without help from a veterinarian. If the fish is lethargic and bloated, it may be worth trying antibiotics - in food if possible.

***Bloated goldfish with cyst***

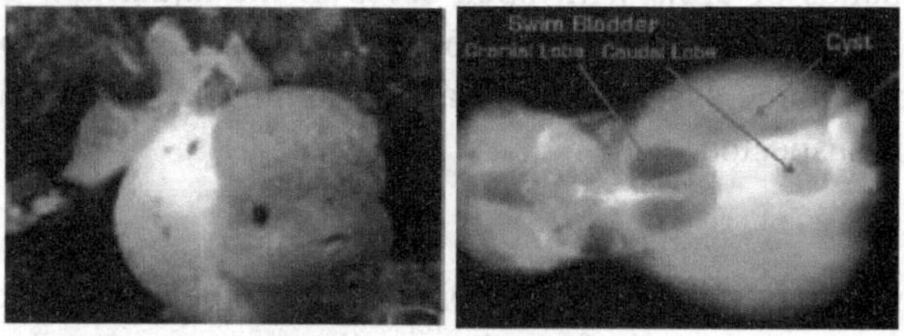


***Dropsy***

Dropsy is a symptom that the fish is in kidney failure and is not able to excrete enough fluid. This generally causes swelling of the belly that is usually accompanied by raised scales that make the fish look like a pine cone. It is difficult to save a fish once this happens, because the kidney is already compromised. Most cases are fatal. Check the gills.

## Causes and Treatments

## Organ failure

Check the gills of the fish. If they look pale, organ failure is likely. The likelihood of saving a fish with organ failure is very small

***Other Causes***

Diseases that can cause dropsy include bacterial, parasitic and viral infections. Toxins in the water or food can also cause kidney failure. The only thing we can do is to provide supportive care and treat as if there is a bacterial cause. If the fish is large enough, antibiotic injections would be the treatment of choice. For smaller fish who are still eating, feed an antibiotic food.

If the fish is not eating and not large enough to inject, use a broad spectrum antibiotic bath. Avoid kanamycin and other aminoglycoside type antibiotics (e.g. amikacin), as they can be nephrotoxic. Many people recommend adding epsom salt (magnesium sulfate) to the water -- we do not believe this helps; however, for hardwater fish, adding 250 ppm calcium chloride may help reduce the energy needed for osmoregulation. Adding salt (sodium chloride) for salt tolerant fish, may also help support osmoregulation.

# 9

# FISH GENETICS

Genetics is the science of genes, heredity, and the variation of organisms. Humans began applying knowledge of genetics in prehistory with the domestication and breeding of plants and animals. In modern research, genetics provides important tools in the investigation of the function of a particular gene, e.g. analysis of genetic interactions. Within organisms, genetic information generally is carried in chromosomes, where it is represented in the chemical structure of particular DNA molecules.

Genes encode the information necessary for synthesizing proteins, which, in turn play a large role in influencing, although, in many instances, do not completely determine, the final phenotype of the organism. The phrase to code for is often used to mean a gene contains the instructions on how to build a particular protein, as in the gene codes for the protein. Note that the "one gene, one protein" concept is now known to be simplistic. For example, a single gene may produce multiple products, depending on how its transcription is regulated.

Areas of genetics Classical genetics Main articles: Classical genetics, Mendelian inheritance

Classical genetics consists of the techniques and methodologies of genetics that predate the advent of molecular biology. After the discovery of the genetic code and such tools of cloning as restriction enzymes, the avenues of investigation open to geneticists were greatly broadened. Some classical genetic ideas have been supplanted with the mechanistic understanding brought by molecular discoveries, but many remain intact and in use, such as the Mendel's laws.

## MOLECULAR GENETICS

Molecular genetics builds upon the foundation of classical genetics but focuses on the structure and function of genes at a molecular level. Molecular genetics employs the methods of both classical genetics (such as hybridization) and molecular biology. It is so-called to differentiate it from other sub fields of genetics such as ecological genetics and population genetics. An important area within molecular genetics is the use of molecular information to determine the patterns of descent, and therefore the correct scientific classification of organisms: this is called molecular systematics. The study of inherited features not strictly associated with changes in the DNA sequence is called epigenetics.

Some take the view that life can be defined, in molecular terms, as the set of strategies which RNA polynucleotides have used and continue to use to perpetuate themselves. This definition grows out of work on the origin of life, specifically the

RNA world hypothesis.

### Population, quantitative and ecological genetics

Population, quantitative and ecological genetics are all very closely related subfields and also build upon classical genetics (supplemented with modern molecular genetics). They are chiefly distinguished by a common theme of studying populations of organisms drawn from nature but differ somewhat in the choice of which aspect of the organism on which they focus. The foundational discipline is population genetics which studies the distribution of and change in allele frequencies of genes under the influence of the four evolutionary forces: natural selection, genetic drift, mutation and migration. It is the theory that attempts to explain such phenomena as adaptation and speciation.

The related subfield of quantitative genetics, which builds on population genetics, aims to predict the response to selection given data on the phenotype and relationships of individuals. A more recent development of quantitative genetics is the analysis of quantitative trait loci. Traits that are under the influence of a large number of genes are known as quantitative traits, and their mapping to a location on the chromosome requires accurate phenotypic, pedigree and marker data from a large number of related individuals.

Ecological genetics again builds upon the basic principles of population genetics but is more explicitly focused on ecological issues. While molecular genetics studies the structure and function of genes at a molecular level, ecological genetics focuses on wild populations of organisms, and attempts to collect data on the ecological aspects of individuals as well as molecular markers from those individuals.

## HISTORY OF GENETICS

The history of genetics started with the work of the Augustinian friar Gregor Johann Mendel. His work on pea plants, published in 1866, described what came to be known as Mendelian Inheritance. In the centuries before-and for several decades after-Mendel's work, a wide variety of theories of heredity proliferated.

1900 marked the "rediscovery of Mendel" by Hugo de Vries, Carl Correns and Erich von Tschermak, and by 1915 the basic principles of Mendelian genetics had been applied to a wide variety of organisms-most notably the fruit fly Drosophila melanogaster. Led by Thomas Hunt Morgan and his fellow "drosophilists", geneticists developed the Mendelian, which was widely accepted by 1925. Alongside experimental work, mathematicians developed the statistical framework of population genetics, bring genetical explanations into the study of evolution.

With the basic patterns of genetic inheritance established, many biologists turned to investigations of the physical nature of the gene. In the 1940s and early 1950s, experiments pointed to DNA as the portion of chromosomes (and perhaps other nucleoproteins) that held genes. A focus on new model organisms such as viruses and bacteria, along with the discovery of the double helical structure of DNA in 1953, marked the transition to the era of molecular genetics.

In the following years, chemists developed techniques for sequencing both nucleic acids and proteins, while others worked out the relationship between the two forms of biological molecules: the genetic code. The regulation of gene expression became a central issue in the 1960s; by the 1970s gene expression could be controlled and manipulated through genetic engineering. In the last decades of the 20th century, many biologists focused on large-scale genetics projects, sequencing entire genomes.

## Ancient theories

The most influential early theories of heredity were that of Hippocrates and Aristotle. Hippocrates' theory (possibly based on the teachings of Anaxagoras) was similar to Darwin's later ideas on pangenesis, involving heredity material that collects from throughout the body. Aristotle suggested instead that the (nonphysical) form-giving principle of an organism was transmitted through semen (which he considered to be a purified form of blood) and the mother's menstrual blood, which interacted in the womb to direct an organism's early development. For both Hippocrates and Aristotle-and nearly all Western scholars through to the late 19th century-the inheritance of acquired characters was a supposedly well-established fact that any adequate theory of heredity had to explain. At the same time, individual species were taken to have a fixed essence; such inherited changes were merely superficial.

In the 9th century CE, the Afro-Arab writer Al-Jahiz considered the effects of the environment on the likelihood of an animal to survive, and first described the struggle for existence.[verification needed] His ideas on the struggle for existence in the Book of Animals have been summarized as follows:

"Animals engage in a struggle for existence; for resources, to avoid being eaten and to breed. Environmental factors influence organisms to develop new characteristics to ensure survival, thus transforming into new species. Animals that survive to breed can pass on their successful characteristics to offspring."

In 1000 CE, the Arab physician, Abu al-Qasim al-Zahrawi (known as Albucasis in the West), wrote the first clear description of haemophilia, a hereditary genetic disorder, in his Al-Tasrif. In this work, he wrote of an Andalusian family whose males died of bleeding after minor injuries.

## Plant systematics and hybridization

In the 18th century, with increased knowledge of plant and animal diversity and the accompanying increased focus on taxonomy, new ideas about heredity began to appear. Linnaeus and others (among them Joseph Gottlieb Kölreuter, Carl Friedrich von Gärtner, and Charles Naudin) conducted extensive experiments with hybridization, especially species hybrids. Species hybridizers described a wide variety of inheritance phenomena, include hybrid sterility and the high variability of back-crosses.

Plant breeders were also developing an array of stable varieties in many important plant species. In the early 19th century, Augustin Sageret established the concept of dominance, recognizing that when some plant varieties are crossed, certain characters (present in one parent) usually appear in the offspring; he also found that some ancestral characters found in neither parent may appear in offspring. However, plant breeders made little attempt to establish a theoretical foundation for their work or to share their knowledge with current work of physiology, although Gartons Agricultural Plant Breeders in England explained their system in their seed catalogue of 1901.

### *Mendel*

In breeding experiments between 1856 and 1865, Gregor Mendel first traced inheritance patterns of certain traits in pea plants and showed that they obeyed simple statistical rules. Although not all features show these patterns of Mendelian inheritance, his work acted as a proof that application of statistics to inheritance could be highly useful. Since that time many more complex forms of inheritance have been demonstrated.

From his statistical analysis Mendel defined a concept that he described as an allele, which was the fundamental unit of heredity. The term allele as Mendel used it is nearly synonymous with the term gene, and now means a specific variant of a particular gene.

Mendel's work was published in 1866 as "Versuche über Pflanzen-Hybriden" (Experiments on Plant Hybridization) in the Verhandlungen des Naturforschenden Vereins zu Brünn (Proceedings of the Natural History Society of Brünn), following two lectures he gave on the work in early 1865.

### *Post-Mendel, pre-re-discovery*

Mendel's work was published in a relatively obscure scientific journal, and it was not given any attention in the scientific community. Instead, discussions about modes of heredity were galvanized by Darwin's theory of evolution by natural selection, in which mechanisms of non-Lamarckian heredity seemed to be required. Darwin's own theory of heredity, pangenesis, did not meet with any large degree of acceptance. A more mathematical version of pangenesis, one which dropped much of Darwin's Lamarckian holdovers, was developed as the "biometrical" school of heredity by Darwin's cousin, Francis Galton. Under Galton and his successor Karl Pearson, the biometrical school attempted to build statistical models for heredity and evolution, with some limited but real success, though the exact methods of heredity were unknown and largely unquestioned.

### *Classical genetics*

The significance of Mendel's work was not understood until early in the twentieth century, after his death, when his research was re-discovered by other scientists working on similar problems. Hugo de Vries, Carl Correns and Erich von Tschermak

There was then a feud between Bateson and Pearson over the hereditary mechanism. Fisher solved this in The Correlation Between Relatives on the Supposition of Mendelian Inheritance

1865 Gregor Mendel's paper, Experiments on Plant Hybridization

1869 Friedrich Miescher discovers a weak acid in the nuclei of white blood cells that today we call DNA

1880-1890 Walther Flemming, Eduard Strasburger, and Edouard van Beneden elucidate chromosome distribution during cell division

1889 Hugo de Vries postulates that "inheritance of specific traits in organisms comes in particles", naming such particles "(pan)genes"

1903 Walter Sutton hypothesizes that chromosomes, which segregate in a Mendelian fashion, are hereditary units

1905 William Bateson coins the term "genetics" in a letter to Adam Sedgwick and at a meeting in 1906

1908 Hardy-Weinberg law derived.

1910 Thomas Hunt Morgan shows that genes reside on chromosomes

1913 Alfred Sturtevant makes the first genetic map of a chromosome

1913 Gene maps show chromosomes containing linear arranged genes

1918 Ronald Fisher publishes "The Correlation Between Relatives on the Supposition of Mendelian Inheritance" the modern synthesis of genetics and evolutionary biology starts. See population genetics.

1928 Frederick Griffith discovers that hereditary material from dead bacteria can be incorporated into live bacteria (see Griffith's experiment)

1931 Crossing over is identified as the cause of recombination

1933 Jean Brachet is able to show that DNA is found in chromosomes and that RNA is present in the cytoplasm of all cells.

1941 Edward Lawrie Tatum and George Wells Beadle show that genes code for proteins; see the original central dogma of genetics

### *The DNA era*

1944 The Avery-MacLeod-McCarty experiment isolates DNA as the genetic material (at that time called transforming principle)

1950 Erwin Chargaff shows that the four nucleotides are not present in nucleic acids in stable proportions, but that some general rules appear to hold (e.g., that the amount of adenine, A, tends to be equal to that of thymine, T).

### *Barbara McClintock discovers transposons in maize*

1952 The Hershey-Chase experiment proves the genetic information of phages (and all other organisms) to be DNA

1953 DNA structure is resolved to be a double helix by James D. Watson and Francis Crick

1956 Joe Hin Tjio and Albert Levan established the correct chromosome number in humans to be 46

1958 The Meselson-Stahl experiment demonstrates that DNA is semiconservatively replicated

1961-1967 Combined efforts of scientists "crack" the genetic code, including Marshall Nirenberg, Har Gobind Khorana, Sydney Brenner & Francis Crick

1964 Howard Temin showed using RNA viruses that the direction of DNA to RNA transcription can be reversed

1970 Restriction enzymes were discovered in studies of a bacterium, Haemophilus influenzae, enabling scientists to cut and paste DNA

### *The genomics era*

1972, Walter Fiers and his team at the Laboratory of Molecular Biology of the University of Ghent (Ghent, Belgium) were the first to determine the sequence of a gene: the gene for bacteriophage MS2 coat protein.

1976, Walter Fiers and his team determine the complete nucleotide-sequence of bacteriophage MS2-RNA

1977 DNA is sequenced for the first time by Fred Sanger, Walter Gilbert, and Allan Maxam working independently. Sanger's lab sequence the entire genome of bacteriophage ?-X174.

1983 Kary Banks Mullis discovers the polymerase chain reaction enabling the easy amplification of DNA

1989 The human gene that encodes the CFTR protein was sequenced by Francis Collins and Lap-Chee Tsui. Defects in this gene cause cystic fibrosis.

1995 The genome of Haemophilus influenzae is the first genome of a free living organism to be sequenced

1996 Saccharomyces cerevisiae is the first eukaryote genome sequence to be released

1998 The first genome sequence for a multicellular eukaryote, Caenorhabditis elegans, is released

2001 First draft sequences of the human genome are released simultaneously by the Human Genome Project and Celera Genomics.

2003 (14 April) Successful completion of Human Genome Project with 99% of the genome sequenced to a 99.99% accuracy.

## Molecular genetics

Although genes were known to exist on chromosomes, chromosomes are composed of both protein and DNA-scientists did not know which of these is responsible for inheritance. In 1928, Frederick Griffith discovered the phenomenon of transformation (see Griffith's experiment): dead bacteria could transfer genetic material to "transform" other still-living bacteria. Sixteen years later, in 1944, Oswald Theodore Avery, Colin McLeod and Maclyn McCarty identified the molecule responsible for transformation as DNA. The Hershey-Chase experiment in 1952 also showed that DNA (rather than protein) is the genetic material of the viruses that infect bacteria, providing further evidence that DNA is the molecule responsible for inheritance.

James D. Watson and Francis Crick determined the structure of DNA in 1953, using the X-ray crystallography work of Rosalind Franklin and Maurice Wilkins that indicated DNA had a helical structure (i.e., shaped like a corkscrew). Their double-helix model had two strands of DNA with the nucleotides pointing inward, each matching a complementary nucleotide on the other strand to form what looks like rungs on a twisted ladder. This structure showed that genetic information exists in the sequence of nucleotides on each strand of DNA. The structure also suggested a simple method for duplication: if the strands are separated, new partner strands can be reconstructed for each based on the sequence of the old strand.

Although the structure of DNA showed how inheritance works, it was still not known how DNA influences the behavior of cells. In the following years, scientists tried to understand how DNA controls the process of protein production. It was discovered that the cell uses DNA as a template to create matching messenger RNA (a molecule with nucleotides, very similar to DNA). The nucleotide sequence of a messenger RNA is used to create an amino acid sequence in protein; this translation between nucleotide and amino acid sequences is known as the genetic code.

With this molecular understanding of inheritance, an explosion of research became possible. One important development was chain-termination DNA sequencing in 1977 by Frederick Sanger. This technology allows scientists to read the nucleotide sequence of a DNA molecule. In 1983, Kary Banks Mullis developed the polymerase chain reaction, providing a quick way to isolate and amplify a specific section of a DNA from a mixture. Through the pooled efforts of the Human Genome Project and the parallel private effort by Celera Genomics, these and other techniques culminated in the sequencing of the human genome in 2003.

## Molecular basis for inheritance

### *DNA and chromosomes*

The molecular basis for genes is deoxyribonucleic acid (DNA). DNA is composed of a chain of nucleotides, of which there are four types: adenine (A), cytosine (C), guanine (G), and thymine (T). Genetic information exists in the sequence of these nucleotides, and genes exist as stretches of sequence along the DNA chain. Viruses are the only exception to this rule-sometimes viruses use the very similar molecule RNA instead of DNA as their genetic material.

DNA normally exists as a double-stranded molecule, coiled into the shape of a double-helix. Each nucleotide in DNA preferentially pairs with its partner nucleotide on the opposite strand: A pairs with T, and C pairs with G. Thus, in its two-stranded form, each strand effectively contains all necessary information, redundant with its partner strand. This structure of DNA is the physical basis for inheritance: DNA replication duplicates the genetic information by splitting the strands and using each strand as a template for synthesis of a new partner strand.

Genes are arranged linearly along long chains of DNA sequence, called chromosomes. In bacteria, each cell usually contains a single circular chromosome, while eukaryotic organisms (including plants and animals) have their DNA arranged in multiple linear chromosomes. These DNA strands are often extremely long; the largest human chromosome, for example, is about 247 million base pairs in length. The DNA of a chromosome is associated with structural proteins that organize, compact, and control access to the DNA, forming a material called chromatin; in eukaryotes, chromatin is usually composed of nucleosomes, segments of DNA wound around cores of histone proteins. The full set of hereditary material in an organism (usually the combined DNA sequences of all chromosomes) is called the genome.

While haploid organisms have only one copy of each chromosome, most animals and many plants are diploid, containing two of each chromosome and thus two copies of every gene. The two alleles for a gene are located on identical loci of sister chromatids, each allele inherited from a different parent.

Many species have so called sex chromosomes. They are special in that they determine the sex of the organism. In humans and many other animals, the Y-chromosome contains the gene that triggers the development of the specifically male characteristics. In evolution, this chromosome has lost most of its content and also most of its genes, while the X chromosome is similar to the other chromosomes and contains many genes. The X and Y chromosomes form a very heterogeneous pair before cell division.

### *Reproduction*

When cells divide, their full genome is copied and each daughter cell inherits one copy. This process, called mitosis, is the simplest form of reproduction and is the basis for asexual reproduction. Asexual reproduction can also occur in multicellular organisms, producing offspring that inherit their genome from a single parent. Offspring that are genetically identical to their parents are called clones.

Eukaryotic organisms often use sexual reproduction to generate offspring that contain a mixture of genetic material inherited from two different parents. The process of sexual reproduction alternates between forms that contain single copies of the genome (haploid) and double copies (diploid). Haploid cells fuse and combine genetic material to create a diploid cell with paired chromosomes. Diploid organisms form haploids by dividing, without replicating their DNA, to create daughter cells that randomly inherit one of each pair of chromosomes. Most animals and many plants are diploid for most of their lifespan, with the haploid form reduced to single cell gametes such as sperm or eggs.

Although they do not use the haploid/diploid method of sexual reproduction, bacteria have many methods of acquiring new genetic information. Some bacteria can undergo conjugation, transferring a small circular piece of DNA to another bacterium. Bacteria can also take up raw DNA fragments found in the environment and integrate them into their genomes, a phenomenon known as transformation.

These processes result in horizontal gene transfer, transmitting fragments of genetic information between organisms that would be otherwise unrelated.

### *Recombination and linkage*

The diploid nature of chromosomes allows for genes on different chromosomes to assort independently during sexual reproduction, recombining to form new combinations of genes. Genes on the same chromosome would theoretically never recombine, however, were it not for the process of chromosomal crossover. During crossover, chromosomes exchange stretches of DNA, effectively shuffling the gene alleles between the chromosomes. This process of chromosomal crossover generally occurs during meiosis, a series of cell divisions that creates haploid cells.

The probability of chromosomal crossover occurring between two given points on the chromosome is related to the distance between the points. For an arbitrarily long distance, the probability of crossover is high enough that the inheritance of the genes is effectively uncorrelated. For genes that are closer together, however, the lower probability of crossover means that the genes demonstrate genetic linkage-alleles for the two genes tend to be inherited together. The amounts of linkage between a series of genes can be combined to form a linear linkage map that roughly describes the arrangement of the genes along the chromosome.

## Gene expression

### *Genetic code*

Genes generally express their functional effect through the production of proteins, which are complex molecules responsible for most functions in the cell. Proteins are chains of amino acids, and the DNA sequence of a gene (through an RNA intermediate) is used to produce a specific protein sequence. This process begins with the production of an RNA molecule with a sequence matching the gene's DNA sequence, a process called transcription.

This messenger RNA molecule is then used to produce a corresponding amino acid sequence through a process called translation. Each group of three nucleotides in the sequence, called a codon, corresponds either to one of the twenty possible amino acids in a protein or an instruction to end the amino acic sequence; this correspondence is called the genetic code. The flow of information is unidirectional: information is transferred from nucleotide sequences into the amino acid sequence of proteins, but it never transfers from protein back into the sequence of DNA-a phenomenon Francis Crick called the central dogma of molecular biology.

The specific sequence of amino acids results in a unique three-dimensional structure for that protein, and the three-dimensional structures of proteins are related to their functions. Some are simple structural molecules, like the fibers formed by the protein collagen. Proteins can bind to other proteins and simple molecules, sometimes acting as enzymes by facilitating chemical reactions within the bound molecules (without changing the structure of the protein itself). Protein structure is dynamic; the protein hemoglobin bends into slightly different forms as it facilitates the capture, transport, and release of oxygen molecules within mammalian blood.

A single nucleotide difference within DNA can cause a change in the amino acid sequence of a protein. Because protein structures are the result of their amino acid sequences, some changes can dramatically change the properties of a protein by destabilizing the structure or changing the surface of the protein in a way that changes its interaction with other proteins and molecules. For example, sickle-cell anemia is a human genetic disease that results from a single base difference within the coding region for the ?-globin section of hemoglobin, causing a single amino acid change that changes hemoglobin's physical properties. Sickle-cell versions of hemoglobin stick to themselves, stacking to form fibers that distort the shape of red blood cells carrying the protein. These sickle-shaped cells no longer flow smoothly through blood vessels, having a tendency to clog or degrade, causing the medical problems associated with this disease.

Some genes are transcribed into RNA but are not translated into protein products-such RNA molecules are called non-coding RNA. In some cases, these products fold into structures which are involved in critical cell functions (e.g. ribosomal RNA and transfer RNA). RNA can also have regulatory effect through hybridization interactions with other RNA molecules (e.g. microRNA).

## Nature versus nurture

Although genes contain all the information an organism uses to function, the environment plays an important role in determining the ultimate phenotype-a phenomenon often referred to as "nature vs. nurture". The phenotype of an organism depends on the interaction of genetics with the environment. One example of this is the case of temperature-sensitive mutations. Often, a single amino acid change within the sequence of a protein does not change its behavior and interactions with other molecules, but it does destabilize the structure. In a high temperature environment, where molecules are moving more quickly and hitting each other, this results in the protein losing its structure and failing to function. In a low temperature environment, however, the protein's structure is stable and it functions normally. This type of mutation is visible in the coat coloration of Siamese cats, where a mutation in an enzyme responsible for pigment production causes it to destabilize and lose function at high temperatures. The protein remains functional in areas of skin that are colder-legs, ears, tail, and face-and so the cat has dark fur at its extremities.

Environment also plays a dramatic role in effects of the human genetic disease phenylketonuria. The mutation that causes phenylketonuria disrupts the ability of the body to break down the amino acid phenylalanine, causing a toxic build-up of an intermediate molecule that, in turn, causes severe symptoms of progressive mental retardation and seizures. If someone with the phenylketonuria mutation follows a strict diet that avoids this amino acid, however, they remain normal and healthy.

A popular method to determine how much role nature and nurture play is to study identical and fraternal twins or siblings of multiple birth. Because identical siblings come from the same zygote they are genetically the same. Fraternal siblings however are as different genetically from one another as normal siblings. By comparing how often the twin of a set has the same disorder between fraternal and identical twins, scientists can see whether there is more of a nature or nurture effect. One famous example of a multiple birth study includes the Genain quadruplets, who were identical quadruplets all diagnosed with schizophrenia.

## Gene regulation

The genome of a given organism contains thousands of genes, but not all these genes need to be active at any given moment. A gene is expressed when it is being transcribed into mRNA (and translated into protein), and there exist many cellular methods of controlling the expression of genes such that proteins are produced only when needed by the cell. Transcription factors are regulatory proteins that bind to the start of genes, either promoting or inhibiting the transcription of the gene. Within the genome of Escherichia coli bacteria, for example, there exists a series of genes necessary for the synthesis of the amino acid tryptophan. However, when tryptophan is already available to the cell, these genes for tryptophan synthesis are no longer needed. The presence of tryptophan directly affects the activity of the genes-tryptophan molecules bind to the tryptophan repressor (a transcription factor), changing the repressor's structure such that the repressor binds to the genes. The tryptophan repressor blocks the transcription and expression of the genes, thereby creating negative feedback regulation of the tryptophan synthesis process.

Differences in gene expression are especially clear within multicellular organisms, where cells all contain the same genome but have very different structures and behaviors due to the expression of different sets of genes. All the cells in a multicellular organism derive from a single cell, differentiating into variant cell types in response to external and intercellular signals and gradually establishing different patterns of gene expression to create different behaviors. As no single gene is responsible for the development of structures within multicellular organisms, these patterns arise from the complex interactions between many cells.

Within eukaryotes there exist structural features of chromatin that influence the transcription of genes, often in the form of modifications to DNA and chromatin that are stably inherited by daughter cells. These features are called "epigenetic" because they exist "on top" of the DNA sequence and retain inheritance from one cell generation to the next. Because of epigenetic features, different cell types grown within the same medium can retain very different properties. Although epigenetic features are generally dynamic over the course of development, some, like the phenomenon of paramutation, have multigenerational inheritance and exist as rare exceptions to the general rule of DNA as the basis for inheritance.

## Genetic change

### *Mutations*

During the process of DNA replication, errors occasionally occur in the polymerization of the second strand. These errors, called mutations, can have an impact on the phenotype of an organism, especially if they occur within the protein coding sequence of a gene. Error rates are usually very low-1 error in every 10-100 million bases-due to the "proofreading" ability of DNA polymerases. (Without proofreading error rates are a thousandfold higher; because many viruses rely on DNA and RNA polymerases that lack proofreading ability, they experience higher mutation rates.) Processes that increase the rate of changes in DNA are called mutagenic: mutagenic chemicals promote errors in DNA replication, often by interfering with the structure of base-pairing, while UV radiation induces mutations by causing damage to the DNA structure. Chemical damage to DNA occurs naturally as well, and cells use DNA repair mechanisms to repair mismatches and breaks in DNA-nevertheless, the repair sometimes fails to return the DNA to its original sequence.

In organisms that use chromosomal crossover to exchange DNA and recombine genes, errors in alignment during meiosis can also cause mutations. Errors in crossover are especially likely when similar sequences cause partner chromosomes to adopt a mistaken alignment; this makes some regions in genomes more prone to mutating in this way. These errors create large structural changes in DNA sequence-duplications, inversions or deletions of entire regions, or the accidental exchanging of whole parts between different chromosomes (called translocation).

## Natural selection and evolution

Mutations alter an organisms genotype and occasionally this causes different phenotypes to appear. Most mutations have little effect on an organism's phenotype, health, or reproductive fitness. Mutations that do have an effect are usually deleterious, but occasionally some can be beneficial. Studies in the fly Drosophila melanogaster suggest that if a mutation changes a protein produced by a gene, about 70 percent of these mutations will be harmful with the remainder being either neutral or weakly beneficial.

Population genetics studies the distribution of genetic differences within populations and how these distributions change over time. Changes in the frequency of an allele in a population are mainly influenced by natural selection, where a given allele provides a selective or reproductive advantage to the organism, as well as other factors such as genetic drift, artificial selection and migration.

Over many generations, the genomes of organisms can change significantly, resulting in the phenomenon of evolution. Selection for beneficial mutations can cause a species to evolve into forms better able to survive in their environment, a process called adaptation. New species are formed through the process of speciation, often caused by geographical separations that prevent populations from exchanging genes with each other. The application of genetic principles to the study of population biology and evolution is referred to as the modern synthesis.

By comparing the homology between different species' genomes it is possible to calculate the evolutionary distance between them and when they may have diverged (called a molecular clock). Genetic comparisons are generally considered a more accurate method of characterizing the relatedness between species than the comparison of phenotypic characteristics. The evolutionary distances between species can be used to form evolutionary trees; these trees represent the common descent and divergence of species over time, although they do not show the transfer of genetic material between unrelated species (known as horizontal gene transfer and most common in bacteria).

## Research and technology

### *Model organisms*

Although geneticists originally studied inheritance in a wide range of organisms, researchers began to specialize in studying the genetics of a particular subset of organisms. The fact that significant research already existed for a given organism would encourage new researchers to choose it for further study, and so eventually a few model organisms became the basis for most genetics research. Common research topics in model organism genetics include the study of gene regulation and the involvement of genes in development and cancer.

Organisms were chosen, in part, for convenience-short generation times and easy genetic manipulation made some organisms popular genetics research tools. Widely used model organisms include the gut bacterium Escherichia coli, the plant Arabidopsis thaliana, baker's yeast (Saccharomyces cerevisiae), the nematode Caenorhabditis elegans, the common fruit fly (Drosophila melanogaster), and the common house mouse (Mus musculus).

### *Medicine*

Medical genetics seeks to understand how genetic variation relates to human health and disease. When searching for an unknown gene that may be involved in a disease, researchers commonly use genetic linkage and genetic pedigree charts to find the location on the genome associated with the disease. At the population level, researchers take advantage of Mendelian randomization to look for locations in the genome that are associated with diseases, a technique especially useful for multigenic traits not clearly defined by a single gene. Once a candidate gene is found, further research is often done on the corresponding gene (called an orthologous gene) in model organisms. In addition to studying genetic diseases, the increased availability of genotyping techniques has led to the field of pharmacogenetics-studying how genotype can affect drug responses.

Individuals differ in their inherited tendency to develop cancer, and cancer is a genetic disease. The process of cancer development in the body is a combination of events. Mutations occasionally occur within cells in the body as they divide. Although these mutations will not be inherited by any offspring, they can affect the behavior of cells, sometimes causing them to grow and divide more frequently. There are biological mechanisms that attempt to stop this process; signals are given to inappropriately dividing cells that should trigger cell death, but sometimes additional mutations occur that cause cells to ignore these messages. An internal process of natural selection occurs within the body and eventually mutations accumulate within cells to promote their own growth, creating a cancerous tumor that grows and invades various tissues of the body.

## Research techniques

DNA can be manipulated in the laboratory. Restriction enzymes are commonly used enzymes that cut DNA at specific sequences, producing predictable fragments of DNA. DNA fragments can be visualized through use of gel electrophoresis, which separates fragments according to their length.

The use of ligation enzymes allows DNA fragments to be connected, and by ligating fragments of DNA together from different sources, researchers can create recombinant DNA. Often associated with genetically modified organisms, recombinant DNA is commonly used in the context of plasmids-short circular DNA fragments with a few genes on them. By inserting plasmids into bacteria and growing those bacteria on plates of agar (to isolate clones of bacteria cells), researchers can clonally amplify the inserted fragment of DNA (a process known as molecular cloning). (Cloning can also refer to the creation of clonal organisms, through various techniques.)

DNA can also be amplified using a procedure called the polymerase chain reaction (PCR). By using specific short sequences of DNA, PCR can isolate and exponentially amplify a targeted region of DNA. Because it can amplify from extremely small amounts of DNA, PCR is also often used to detect the presence of specific DNA sequences.

## DNA sequencing and genomics

One of the most fundamental technologies developed to study genetics, DNA sequencing allows researchers to determine the sequence of nucleotides in DNA fragments. Developed in 1977 by Frederick Sanger and coworkers, chain-termination sequencing is now routinely used to sequence DNA fragments. With this technology, researchers have been able to study the molecular sequences associated with many human diseases.

As sequencing has become less expensive, researchers have sequenced the genomes of many organisms, using computational tools to stitch together the sequences of many different fragments (a process called genome assembly). These technologies were used to sequence the human genome, leading to the completion of the Human Genome Project in 2003. New high-throughput sequencing technologies are dramatically lowering the cost of DNA sequencing, with many researchers hoping to bring the cost of resequencing a human genome down to a thousand dollars.

The large amount of sequence data available has created the field of genomics, research that uses computational tools to search for and analyze patterns in the full genomes of organisms. Genomics can also be considered a subfield of bioinformatics, which uses computational approaches to analyze large sets of biological data.

## GENETICS IN FISHERIES MANAGEMENT

The role that the application of genetic principles can play in the sustainable use and conservation of living aquatic resources is being increasingly appreciated by resource managers, policy makers and the international community. Gene and genotype frequency data can provide information on inter alia

- species identification,
- population stock structure,
- hybridization,
- gene flow,
- inbreeding,
- and the genetic impacts of fishing and aquaculture.

Often, data from other sources, e.g. tag-recapture and morphometrics, cannot provide such information or are extremely difficult to collect in areas such as the Mekong River Basin. Genetic data and other types of data should be used together to understand better the dynamics of aquatic populations both in culture and in the wild.

For species such as Pacific salmon that home to natal streams to reproduce, genetic stock identification is a cost effective means to manage fisheries based on genetic differences among stocks of salmon.

Many other groups of fishes and invertebrates show similar genetic structuring because of geological changes, current patterns, among other things. For many marine species the differences among sub-groups is not large because of spawning behavior, migrations and not as accurate homing to natal areas as salmon. Genetic differences have also been used to give endangered species protection to valueable groups of fishes for example in the USA under the Endangered Species Act. Application of genetic improvement in aquaculture and culture-based fisheries is progressing rapidly and population genetics can also be applied to the study of cultured groups of fish in order to choose appropriate breeds to crossbreed and to monitor inbreeding.

### Molecular Ecology and Fisheries Genetics Laboratory

The Molecular Ecology and Fisheries Genetics Laboratory (MEFGL) is a leading research division within the School of Biological Sciences' (SBS) Molecular Ecology and Evolution Group, one of the largest research areas within the University. The MEFGL represents one of Europe's largest centres focusing on population and species diversity of aquatic animals, with additional activities on terrestrial mammals and invertebrates. A central part of our work is the application of molecular markers such as microsatellites, AFLPs, single nucleotide polymorphisms (SNPs) and sequence data (454 Roche and Sanger) to fundamental evolutionary and ecological questions relating to the origins, levels, distribution and ecological significance of genetic variation in wild populations. In addition to the focus on neutral markers, facilities and expertise are expanding on the functional analysis of genomes, including application of transcriptomics and Real-Time PCR. The MEFGL complements molecular genetic data with detailed ecological, behavioural and physiological information to facilitate examination of the underlying causes of individual, population or species diversity in time and space. Summary details of MEFGL activities, scope and facilities can be obtained from the downloadable MEFGL Research Pamphlet.

## Comparative Testing of Fish Species and Different Lines

The aim of comparative testing is to determine and compare the most important characteristics which are closely correlated with the productivity of species. The different phases of comparative testing are as follows:

1. The groups for testing have to be produced with artificial propagation. One female and one male can be used which are chosen by phenotypical selection with the best growth rate and appearance.
2. The groups of progenies will be reared in separate ponds and after six months they need to be stocked together in a common pond with group marking (150-200 fish from each population at least). The stocking rate of testing ponds must be the same as in commercial fish farming/intensive polyculture. When the number of stocked fish from one population are less than 100 fish, the testing work has to be repeated three to four times with the same groups of fish.
3. It is necessary to use organic and inorganic fertilizers to fertilize the water and also to feed the fish.
4. The fish population reaching a marketable size need to be harvested. Proper measurement of every individual is required for evaluating different characteristics.
5. The qualitative characteristics, such as the colour, scales and the formation of fins are not closely correlated with the productivity, therefore the evaluation of those only for the purpose of morphological characterization of the populations has any importance.
6. From the quantitative characteristics one has to evaluate the viability, growth rate, food conversion, quantity of meat, quality of meat (protein and fat content) and the reaction to fish diseases. Later, by producing inbred lines, the fecundity and the oxygen demand tolerance also has to be examined.
7. The results obtained need to be verified with variance analysis.
8. The productivity of various groups can be compared through a multifactorial,

"100 score" system by the four most important characteristics (like viability, growth rate, food conversion, quality of meat).

9. If the test is arranged in more ponds with replicates, standard population control has to be used in all ponds, which gives the possibility to compare the differences of environment among the ponds used. The common carp is suggested as a standard control considering its omnivorous feeding habits. The standard control can be produced in every year with a panmictic propagation, maintaining their quality and characteristics on the same level during a long time.
10. The results of pond testing experiments can be compared with the short laboratory tests and with the results of oxygen demand toleration (ODT) tests. The average of the above three methods gives a reliable final result of comparative testings.
11. The outstanding populations showing best characteristics will be used for the further selection and hybridization work.

Progeny Testing for Determining the Individual Breeding Value

The progeny testing is a suitable method to determine the individual breeding value of males and females after the productivity of their progenies.

***The phases of the work are the following:***

1. For the evaluation of males, the eggs of two females have to be fertilized with the sperm of chosen males separately, producing maternal half sister groups. Five or six males can be examined at the same time in their ten or twelve progeny groups from the two different females. For the evaluation of females one must follow the same system, using two different males.
2. The productivity of half sister groups are examined and compared by using the above mentioned comparative testing method (i.e., 4.1.1).
3. The average productivity of two groups originating from different females but from the same male shows the individual breeding value of the given male fish.
4. The best males and females in breeding value can be used for the intraspecific hybridization, or for improving the species in their characteristics.

## Intraspecific Hybridization

The aim of intraspecific hybridization is to produce more productive hybrids by crossing different genotypes of females and males having the best results during the previous progeny testing. The phases of intraspecific hybridization are as follows:

1. Selection of various genotypes of the same species;
2. Inbreeding of the selected lines;
3. Polyallel crossing of the different lines;
4. Comparative testing of the F1 hybrid generation;
5. Propagation of the best hybrid producing parental lines for the fish seed production farms;
6. Producing three-line hybrids;
7. Producing four-line hybrids.

After the inbreeding of parental lines, the heterosis effect can be increased. The fastest method of inbreeding is the artificial gynogenesis.

As a result of intraspecific hybridization, the increasing productivity of fish species of about 15-25 percent can be expected.

## Artificial Gynogenesis and Hormonal Sex Reversion

The main purpose of gynogenesis is to produce inbred individuals having only maternal chromosomes for the further genetic work. For causing gynogenesis artificially, inactivated sperm can be used for fertilization of eggs, with a cold shock treatment after fertilization. For the inactivation of sperm a cobalt 60 gamma ray source or ultra-violet germicide lamp has to be used. The individuals of gynogenetic populations are females with diploid chromosome sets similar to the mother. The technique of artificial gynogenesis was demonstrated on the common carp. The method can be adopted for the other fish species with necessary modifications.

For producing gynogenetic inbred male populations, methyl-testosteron hormone feeding has to be provided for the females at a young age to influence the sex formation in the direction of the male. The technique of sex reversion was demonstrated with gynogenetic common carp. The same method can be used for all fish species which are able to feed with artificial food.

The inbreeding coefficient value (F) of the first gynogenetic population is already 0.90, which value can be reached only after 12 generations of sister-brother pairing.

For the purpose of increasing the inbreeding level, the gynogenesis can be applied from generation to generation. It is important to make a proper selection within the gynogenetic population to avoid the undesirable effect of inbreeding depression. The keeping and maintenance of gynogenetic inbred females and males requires more care than the normal ones because of their high breeding value and more sensitiveness to the environmental effects. Therefore, the location, rearing and keeping of this fish in small ponds is necessary, with a good water supply and other necessary conditions and to protect them from poaching. Some outstanding individuals can be kept in an indoor raceway system.

The gynogenetic inbred populations or individuals can be used for improving the species with their useful characteristics, to make inbred lines for intraspecific hybridization, for producing monosex fish populations and also for genetic examinations for particular purposes.

## Intergeneric Hybridization

Intergeneric hybridization consists of crossing fish belonging to different species. The aim of producing intergeneric hybrids is as follows:

1. To unite the useful characteristics of two different species in one of the new forms.

2. Producing a new construction in consequence of the heterosis effect.

Using intergeneric hybridization, the following results can be expected:

1. New construction with new feeding habits for composite fish culture or for water reservoirs
2. Better growth rates and survival rates
3. More restful behaviour, having better tolerance for reduced levels of oxygen (silver carp × bighead)
4. Better quality of meat (common carp × silver carp; rohu × catla)
5. Monosex triploid grass feeder fish (grass carp × bighead)
6. Better tolerance of water salinity for brackishwater fish culture.

In the field of intergeneric hybridization the scientists of India have achieved good results (see Section 3).

The phases of intergeneric hybridization are the following:

1. Studying the fish species characteristics; behaviour and genetic structures
2. Analysis of chromosome sets and karyotypes
3. Crossing the species, using artificial propagation
4. Rearing the hybrids in the most suitable environment for their expected habitat
5. Comparative testing of hybrids and the parental forms
6. Selection of the most suitable individuals
7. Artificial propagation experiments on the F1 hybrids
8. Stabilizing the new characteristics with inbreeding or back-crossing with the closer parental species which are similar in their main characteristics to the new hybrids
9. Describing the new hybrids' morphological, biological and economic characteristics.

Application of Gynogenesis for Improving Productivity of Fish Species

Producing and maintaining inbred lines for intraspecific hybridization

The greatest advantage of producing inbred lines is to obtain higher heterosis effect after the crossing of gynogenetic female and male lines.

The main phases of the work are as follows:

1. To select the best females for performance and morphological features.
2. Producing the first gynogenetic female generation (G1)
3. Rearing the gynogenetic fry suitably for producing sexually mature fish as soon as possible.
4. Selection by phenotype for the best growing and correct body formation.
5. Producing the second gynogenetic female generation (G2).
6. After dividing the G2 population into two parts, one part will be reared as a female group and the other part has to be oversexed for producing gynogenetic males (G2 ).
7. Rearing the G2? and G2 groups suitably.
8. Brother-sister pairing can be used for producing the third gynogenetic generation (G2? × G2 - G3?). The purpose of brother-sister pairing is to avoid the chance injuring of chromosomes.

9. Sex reversion in the G3? progeny group.
10. Brother-sister pairing for producing fourth gynogenetic generation (G2? × G3 - G4)

This scheme can be continued by the previous pattern. During all the phases the best female or male are selected according to their better constitution.

The inbred population produced can be used after phase five for intraspecific hybridization. After top crossing, good heterosis effect can be expected.

## Producing semi-gynogenetic lines

The purpose of this work is to produce gynogenetic inbred male lines in cases when the hormone feeding is different, such as the case of silver carp and catla. The phases of work are as follows:

1. From population "A" selection to be made of the best females.
2. Producing the first gynogenetic generation (AG1?).
3. Crossing the AG1? female with a normal A? male. The result obtained will be a semi-gynogenetic bisexual population (50% ? and 50% ?).
4. Selection for the best females and males.
5. Producing the second gynogenetic generation (AG2?)
6. Crossing of the AG2? female with the normal A? male. The scheme can be continued by the previous pattern. After the third phase the gene pool of population becomes homozygotic to about 60 percent.

### *Producing monosex fish populations with gynogenesis*

The purpose of this work is: (i) to produce monosex female populations for pond fish culture because of their faster growth than that of the bisexual group of the same age; (ii) the stocking of monosex populations in natural waters like reservoirs or undrainable lakes; (iii) to develop inbred gynogenetic male lines is necessary for maintenance only in commercial fish producing farms or for fish seed production farms and pairing them with normal females; (iv) after solving the problem of deep freezing the sperm, one can collect and store the sperm of inbred gynogenetic suitable males, so as to use it for producing monosex populations in other farms.

The phases of the work are as follows:

1. To select female from a male line
2. Producing the first gynogenetic population (G1 ?)
3. Producing a gynogenetic male population with hormonal sex reversion (G1 )
4. Using the selected males for top-crossing by pairing them with normal females.

The progenies will be all females which may show improved characteristics.

# 10

# FISH MEDICINES & TREATMENTS

When dealing with fish diseases, first you have to identify what you're treating. There is no mediation that is effective against everything. Few fish medications are even good at treating a couple different pathogens. Get a good book on fish diseases and study it. Even then, you will most often have to make your best guess as to what fish medication to use.

Also realize that it is more common than not, that you will be dealing with multiple disease problems. Most diseased fish are so weak that their immune system will not prevent attack from other opportunistic pathogens.

## How to Prepare

Before deciding to treat, first check the basic water parameters like temperature, pH, hardness, ammonia and nitrite levels. You don't want to stress fish with a medicine if the problem is something other than a pathogen. Also check to see that the filter is not clogged or in need of cleaning. Be sure there is good water circulation which is necessary for optimum oxygen levels. Look for fish that are bullying others.

There may also be an ornament or gravel that is leaching a toxin. Make sure they're aquarium safe. Before treating, do a water change. This is often the most effective treatment and sometimes the problem goes away with increased water changes alone.

Deciding what to buy after you have a problem almost never works. Quick action is almost always very important, therefore, have some medications on hand and have a system set up so you can implement the course of action quickly.

First, have a hospital tank (or two) ready with a biologically active filter. Use this specifically for treating fish and nothing else. Use the smallest tank appropriate for the fish you're treating. It will save on medication, make the frequent water changes easier and is less of a pain to set up and take down.

Several medications will stain the silicone, so there is no sense using show tanks to medicate fish in. Malachite green, methylene blue and copper sulfate are some of the worst at leaving stains. Many medications will also kill plants and invertebrates. A hospital tank will treat only the fish.

The hospital tank should have direct light over it. This keeps the fish more calm and bacteria grow faster under bright light.

Light will also degrade some medications. When done with the treatment, sterilize the tank and filter for the next time you need it. After sterilization, the filter can be added to a healthy tank to re-seed the bacteria for the next use.

Many medications will harm the nitrifying bacteria in your filter, so have an extra filter in healthy tank, ready to be put into the hospital tank if needed. Nitrifying bacteria are gram positive aerobic bacteria, so if you use a medication like erythromycin, the filter will be killed.

Some broad spectrum antibiotics like Kanamycin Sulfate are more effective against gram negative bacteria, but will kill some gram positive aerobic bacteria, like nitrifying bacteria, when used heavily.

Medicated foods will protect your filter, but when using a medicated food, it is critical that the fish are accustomed to eating this same exact food without the medications. Medicated food only works if the fish consume a particular amount of food. In general, the medicated foods are formulated for a fish that is eating 2-3% of it's body weight in food each day.

Treatment Guidelines: If you suspect an internal parasite - high heat is beneficial. The parasites don't like it and some are actually killed by it. In addition, the fish's immune system is boosted at higher temps. 95-96°F is the most effective heat range. Most tropical fish can easily handle this if the tank is well aerated and the temperature is raised gradually over a full one day period. If in doubt, simply carefully monitor them while raising the temperature.

If you think it's a bacterial problem, high temperature is harmful. Bacterial thrive in the 1980's and 1990's, so it's best to treat in the low 1970's. At this temperature, they won't eat much, so feed very lightly. In fact, if you suspect an internal bacterial problem, food is not not going to help. Unless it's an antibiotic food, the fish are better off not eating. If fish do not respond to the antibiotic within four days, discontinue it and try another. If they do respond, do not stop treatment. The full course must be given or they will be re-infected and become worse than before. Be aware—many manufacturers recommend a dose that is inadequate for a cure and many also have "cut" formulas that have a bulk inert ingredient added, which further dilutes the effectiveness of the medication. Pay attention to the purity of the medication.

Always choose a medicated food for an internal problem if you can find one with the proper medication and the fish is eating well. It is more effective, less harmful to the environment, saves the bio filter and is a less expensive way to treat. You can make your own medicated foods by combining the food and medication with gelatin to hold it together. Just soaking a dry food in medication is not very effective. The medication dissolves into the tanks water within seconds and the dose cannot be controlled.

Expired medications: Most medications will retain over 90% of their effectiveness well past their expiration date. The dates mandated by the government are more than just a little conservative. There are a couple, like Trichlorfon (dylox) and Oxytetracycline Hydrochloride that can become toxic when they get too old, but even they are probably good for well over a year past the expiration date. Keep them cool and dry for maximum shelf life and effectiveness.

Overview of common bacterial types:

| **Gram Positive** | **Gram Negative** |
|---|---|
| Mycobacterium (TB) | Aeronomas - anaerobic |
| Streptococcus - aerobic | Furunculosis - anaerobic |
| Pseudonocardio - anaerobic | Vibro - anaerobic |
| Staphylococcus - aerobic | Columnaris - aerobic |
| Cynobacteria (blue/green algae -phototrophic | Pseudomonas - aerobic |
| | Strptobacillus - anaerobic |
| Nitrifying - aerobic | Salmonella - anaerobic |

## Methods of Disease Treatment

Diseased fish should be isolated whenever possible so they are not likely to spread infections among other healthy fish in the tank. The best way to isolate sick fish is to set up a hospital tank.

### *Hospital Tank*

The hospital tank should have no substrate or live plants. Provide shelter with over-turned flower pots and use plastic plants to help comfort the fish. Use a sponge filter or internal box filter that does not use chemical filtration or Carbon, for these products will remove many types of medications.

It is not recommended to use a strong power filter as the water disturbance may aggravate ailing fish. The tank should be furnished with a reliable heater that will not permit much temperature variance. To protect fish from heater burns, the heater should be enclosed in plastic mesh or other other means of coverage. The tank should be dimly lit and kept in a dark room Some pathogens do poorly in absence of light. Lastly, the tank should be furnished with an air stone to provide oxygen as some medications reduce the amount of free oxygen in the water.

When adding the fish to the hospital tank, try to have the water conditions as close to those from where the fish is coming. The initial stress is reduced when conditions are similar. After treating the sick fish, disinfect the tank with dilute hydrochloric acid or potassium permanganate. Let the chemical stand in the water for two or three days and then thoroughly clean the tank.

After disinfection, the biological filtration should be restored, so the tank will be ready for the next sick fish. The hospital tank can double as a quarantine tank. Ideally, all new fish should be quarantined for three to six weeks before they are added to the main tank.

### *Baths*

An alternative means of treating a fish is to give it a short "bath" in a bucket containing water and a medication or salt. Chemicals for such baths, such as Formalin, can be found in aquarium stores. A salt bath can be used for a wide range of disorders. The length and concentration of the bath can be varied.

### *Heat Therapy*

Heat therapy of raising the water temperature, can be administered to kill certain parasites. Heat therapy should only be used in clean tanks with an abundant supply of oxygen.

The temperature should be raised gradually by 1.8°F (1°C) an hour until the desired temperature is reached. During this time, the fish's condition must be monitored, because sometimes stressed fish cannot handle such a temperature change. Discontinue treatment if any side effects occur.

### *Medicated Foods*

One means of combating internal parasites is to use a medicated fish food.

### *Commercial Fish Medications*

Commercial fish medications are not regulated by any government agency. This means that some commercial medications have not been tested for safety or effectiveness. Thus some medications are completely useless and may be even harmful to aquarium inhabitants. It is suggested that the hobbyist only purchase medication with the ingredients, and their concentrations, printed on the label.

A list of ingredients only is as good as the aquariasts knowledge of the ingredients' effectiveness, so look for compounds listed in this, and other books. Well-known brands (i.e. Tetra) are usually safe bets, but one can never be sure unless the ingredients are available.

### *Killing a Fish*

Sometimes killing a fish suffering from an incurable disease is more humane than allowing it to continue suffering. There are several acceptable ways to kill a fish: severing the spinal cord just below the head with a sharp pair of scissors; placing the fish in water and rapidly cooling the container with ice; or overdoing with an anesthetic.

### *Being Prepared for Disease*

It is always a good idea to be prepared for an outbreak of disease. Thus one should keep all the supplies, not necessarily a number of medications, that they would need for disease treatment. Among the materials that should be kept are a hospital tank, buckets, and salt.

The aquariast should also have knowledge of their fish's normal behavior—when they feed, how they swim, what they look like, when their active—so that they can know when the normal behavior has changed.

### *Steps to Treat Ailing Fish*

If a fish is ailing, there are several steps that should be taken prior to using medication.

(1) Take a look at the ailing specimen, and take note of what is different from normal.

(2) Check to make sure that all environmental conditions are in order, because the number one cause of stress, ailing behavior, disease, and death in aquaria are environmental problems. Test the pH, ammonia, nitrite, and nitrate levels, and if possible, the oxygen content. If these are in order, make sure that no other toxins such as paint fumes, pesticides, or cleaning agents could have entered the tank. If the answer is no again, make sure all accessories are operating properly (heater, air pumps, filter, etc.).

Try to see if another tank mate may have been on the aggressive. Run down the list of the fish's requirements to see that they are all in order. Remember that tank layout and dietary deficiencies also can affect the fish's health.

If you find that one these conditions is off, then take steps to fix it. If the water chemistry is the problem, make water changes until the problem is fixed. If everything checks out, it is time to move on to the next step.

(3) See that the hospital tank is in order, and set up two buckets full of water (The water should match the main aquarium's water conditions and be dechlorinated.). The buckets should be at least 2.5 gallons (9.5 L), though 10 gallon (38 L) tanks are preferable for large fish. Be sure that the containers are well-aerated.

(4) Move the ailing fish by net to the first bucket. Add 3/8 TSP of salt per gallon (0.5 g/L) every two minutes for 10 minutes. Using the net, move the fish to the next bucket of fresh water for 30-45 minutes. After 15-20 minutes, add three-eighth TSP of salt per gallon (0.5 g/L) again.

(5. Net the fish and move it into the hospital tank. Be sure to change the water daily (30-50%). After a day, move the fish into a shallow, clear bowl. Examine the fish for parasites and evidence of disease. If necessary, use a magnifying glass.

## AQUARIUM MEDICATIONS & TREATMENTS
## TRICHLORFON (DYLOX)

Dimethyl (2,2,2,Trichloro-1-Hydroxyethyl) Phosphonate: Trichlorfon is an Organophosphate and degrades rapidly (approximately 99% of applied degraded in 2 hours) in alkaline pond water (pH 8.5) at room temperature. But remains stable in the same pond water held under acidic (pH 5.0) conditions for two hours.

Trichlorfon is very highly toxic to invertebrates.

Do Not Use on Fish that are Chemically Sensitive such as: Silver Dollars, Rays, Bala Sharks, Arowanas, Tinfoil Barbs, Hemodias, Piranha, Most Silver Scaled Fish, Marine sharks, Lion Fish.

See methylene blue if Dylox treatment is not available or viable.

Use: Trichlorfon is useful for treatment of: Hydra, Lernia (Anchor Worms), Parasitic Copepods, Monodigenetic and Digenetic Flukes, Fish Lice (Argulus), Leeches.

Clout contains Trichlorfon (Dimethyl (2,2,2,Trichloro-1-Hydroxyethyl) Phosphonate), as well as; 4-[P-(dimethylamino)-O-phenylbenzylidene]-2,5-cyclohexadien-1-xylidene dimethylammonium chloride, 1,2,dimethyl-5-nitroimidazole.

Dosage: There is no recommended dosage (other than Freshwater fish acute toxicity= 1.6-180 ppm), refer to manufacturers directions for all products containing Trichlorfon.

Treat again in fourteen days to kill new hatchings of Anchor worms. Use of Trichlorfon for snail treatment; caution be exercised when Trichlorfon is used for the removal of snails (or if snails are present during a parasite treatment). Although I will admit the evidence at this point is anecdotal, but since Trichlorfon is highly toxic to snails, the use of this treatment to kill snails (or in their presence) can rapidly decrease pH (especially in tanks in low KH) as snails rapidly die off (which can release other toxins).

An important fact may be "at play" in that Trichlorfon is more stable in lower pH water, thus retaining its toxicity for a longer period of time which the often high organic decomposition found in tanks with high snail populations may allow a "snow ball" effect of dying snails, increasingly lower pH, and increasing toxicity of Trichlorfon which may result in the death of fish as well.

## INTERNAL PARASITE MEDICATIONS

### Piperazine

It is an organic compound used as an anti-parasitic in veterinary medicine, primarily for worms. Piperazine works through anthelmintic action (used to expel or destroy parasitic worms in the gastro-intestinal tract). Their mode of action is generally by paralysing parasites, which allows the host body to easily remove or expel the invading organism. This action is mediated by its agonist effects upon the inhibitory GABA (the chief inhibitory neurotransmitter in the vertebrate central nervous system.) receptor.

Its selectivity for worms and similar invertebrates is because vertebrates only use GABA in the Central Nervous System and a worms GABA receptor is a different isoform (A protein having the same function and similar or identical sequence, but the product of a different gene and usually) to the vertebrate's one.

This product should never be used in the presence of invertebrates as for the reasons outlined above. Piperazine aquatic uses are restricted to internal parasite control, especially intestinal worms. Piperazine is proven effective for Capillaria Nematode worms that infest the intestines of Angelfish, Discus, some other cichlids, and occasionally other aquarium species. In the aquarium the disease spreads easily from fish to fish as they consume the eggs of the worms, shed in feces of infected individuals.

Piperazine is found in Prazipro, Pepso Flakes, and the now discontinued Tetra Anti-Parasite Flakes.

## Praziquantel

It is an anti-worm, medication. It prevents worms from growing or multiplying in the body. It is used to treat infections caused by worms. Also internal trematodes, commonly known as flukes and flatworms. This product can be used for external Flukes when safety of more effective treatments such as Trichlorfon are in question.

Praziquantel is found in Jungle Internal Parasite Guard and Aquarium Products Gel Tek Medicated Food which is a better mode of treating with this medication.

Praziquantel and Levamisol are found in jungle medicated parasite food which is an excellent way to deliver this food.

## Levamisol

It has a wide range of anti-nematodic (worms) effects. It is efficient in destroying both adult and larvae forms of gastrointestinal and pulmonary nematode parasites in fish (& is one of the best nematode treatments one can use). It attacks parasites by causing first paralysis and then death of the parasites. Partly decomposed parasites are being eliminated during the first twenty four hours since the beginning of the medicine application.

Levamisole is found primarily in wormers for animals at veterinary supply houses and treated a 5 ppm. This was also found in an Aquatronics product no longer available (a good one at that); Discomed.

Dosage: Refer to product instructions containing these medications Do not use any of these in the presence of invertebrates. If Levamisol cannot be loacted, the use of Levamisol HCL found in many commercial dog, poultry, cattle, etc. wormers can be substituted.

fifty nine mg of Levamisole HCl is equivalent to 50 mg of pure levamisole.

You will 2.36 mg/L (or 9 mg/Gallon) of Levamisole HCL, so approximately 90mg of Levamisole HCL will treat 10 gallons (38 liters) with a required 2 ppm concentration. Since Levamisole can be safely over dosed (with a considerable safety margin), approximately one-fourth teaspoon will work per 10 gallons (.19 teaspoons is more exact).

## METHYLENE BLUE (Zinc Free) 2.303%

A heterocyclic aromatic chemical compound with molecular formula: $C16H18ClN3S$.

Methylene Blue is widely used a Redox indicator in chemistry. Solutions of this substance are blue when in an oxidizing environment, but will turn colorless if exposed to a reducing agent.

Since Methylene blue is a redox dye and raises the oxygen consumption of cells, this causes the hydrogen oxidized to be passed on to the oxygen. Each molecule of the dye is oxidized and reduced about hundred times per seconds. Thus, while disinfection results from this, methylene blue is also excellent against methemoglobin intoxication.

The therapeutic action of methylene blue on bacteria and other parasites is probably due to its binding effect with cytoplasmic structures within the cell and also its interference with oxidation-reduction processes.

Also due to its oxidative reduction properties, MB can also be used as an indicator to determine if a cell alive or not or if the slime coat which is essential for fish is healthy or not. The blue indicator turns colorless in the presence of most healthy cells, slime coat, or active enzymes, HOWEVER the fish will stain blue where injury has occurred, especially to the slime coat (which the loss of would compare to a human have skin peeled off).

Methylene blue is used as a medication for the treatment of methemoglobinemia. Methemoglobinemia can be caused by high nitrites (and ammonia) in the blood, which happens in fish respiration in water high in ammonia or nitrites. Methemoglobinemia is treated with the use of methylene blue, which restores the hemoglobin to its normal oxygen-carrying state.

Methylene Blue is made from the anti-malarial drug, chloroquine. Malaria is a protozoan similar to the protozoan that causes ich (Ichthyophthirius multifilius) and oodinium, which is why Methylene Blue is useful for parasite and fungal treatments.

This is not to advocate for the use of Methylene Blue in this way), it has now been shown to help Alzheimer's patients slow the progression of Alzheimer's disease for as long as nineteenth months used in 60 mg. doses three times daily.

For use to treat Fungus on eggs, Ich, Saprolegnia and some bacteria (although generally not for use in bacterial infections other than as a bath/dip/quarantine therapy).

MB is effective in treatment of some gill diseases, especially from injury such as ammonia burns or fluke damage as it transports oxygen allowing for more gill efficiency and as well aiding in some healing of tissue.

Probably one of the best "first response" treatments when used as a bath or even dripped directly on a fish for both internal and internal injuries and infections. Generally used as a thirty minute bath/dip at double dose, in fact Methylene Blue has a wide safety margin and is non-toxic when used as recommended. Fish tolerate relatively high dosages of Methylene Blue without side-effects.

Very effective when used as a dip for topical treatment of parasites, fungal, and some infections. When Dylox is not available, this can a useful treatment for anchor worm especially in goldfish; first carefully remove the anchor worm with tweezers, then dip the affected fish in Methylene blue. When used with a ultraviolet sterilizer to kill the swimming stage of the female anchor worm (the female is the parasite), this can be an effective treatment. Tank can also be treated with Malachite green or malachite green combination during this time for improved effectiveness.

Methylene blue can also treat some protozoa such as Oodinium, although only a mild treatment for this but it can be more effective when combined with other chemical treatments such as in Medicated Wonder Shells.

Also very useful for;

A dip/bath for potassium cyanide, ammonia, and nitrite poisoning due to Methylene Blue's affect on Methemo globinemia (nitrite poisoning).

Effective as an antidote for other forms of poisoning including damage to the liver and kidneys caused from poisoning (assuming damage is not past the point normal regeneration) due to being reduced by components of the electron transport chain (a chemical reaction between an electron donor and an electron acceptor to the transfer of H+ ions across a membrane, via a set of mediating biochemical Redox reactions).

For transfer of fish when moving or temporary storage of fish in crowded conditions.

Treatment of new fish arrivals in a hospital tank, again due to methylene blue's affect on Methemoglobinemia, bacteria, and protozoa.

As a medicated bath for Dropsy or any other internal malady such as Swim Bladder problems (as MB is easily tissue absorbed). For many external infections Potassium Permanganate is sometimes a better choice for a bath.

A test for Redox Potential in aquariums, as methylene blue will lose color in a reducing environment (you can test this with some powdered Wonder Shell dissolved in water).

Dosage: 1 teaspoon of a 2.303% solution per 10 gallons every other day for ten days with water changes before each treatment. Best Used in a Hospital Tank. Methylene blue can destroy nitrifying bacteria and plants in the display aquarium.

Many persons preferred use of Methylene Blue is a bath. To prepare this bath use 1 teaspoon 2.303% solution per 5 gallons (double dose) in a bath of aquarium water from the tank the fish you wish to treat came from, we usually use about a ½ gallon of water, however you may use less. Measurement of the Methylene Blue does not need to be precise as this bath should be used for about thirty minutes (although do not overdose). One may also add salts to his bath to improve effectiveness for certain problems when preparing baths for freshwater fish (such as swim bladder, dropsy or unknown problems). Generally, we would use one teaspoon of sodium chloride (regular salt) per gallon and occasionally ¼ teaspoon of Epsom Salts (Magnesium Sulfate) per gallon.

Make sure one keep the water in a warm area, as in a cold room the water temperature can drop rapidly which would stress the fish. Do not pour this water back into your display aquarium when finished. This can be performed twice per day. This bath is useful for fresh and saltwater.

## Aquarium Disease Prevention; Fish Baths

This bath is very effective for ammonia/nitrite poisoning, Swim Bladder problems, and is helpful for many other internal/systemic infections. MB baths are a good treatment when nothing else is working and/or little is known as to why as fish is acting abnormally (the MB bath does not always work, but it is a relatively safe method vs. indiscriminately dumping other medications into a display aquarium).

In the case of swim bladder infection or problems the MB bath is often the only treatment required other than possible correcting diet and adding more electrolytes (calcium, magnesium, etc. to your aquarium.

It is important to note that you Methylene Blue is removed by activated carbon filtration. It will also be absorbed by porous materials such as rock, coral and wood. Caution should be used in a display tank at full dose (not usually a problem when part of buffered chemical formulations such as Medicated Wonder Shells when used as directed).

## FORMALIN (3% formaldehyde):

Use: For treatment and control of the diseases caused by protozoan and monogenetic trematodes of freshwater and marine aquarium fishes. Formalin will control or help prevent diseases of fishes caused by the following disease organisms: Ichthyophthirius (freshwater "ich"), Costia (Ichthyobodo), Chilodonella, Ambiphyra, Cryptocaryon (marine "ich"), Epistylis, Oodinium, Amyloodinium, and Trichodina. Formalin is also often effective against the common external fungal infections of fishes and their eggs caused by fungus (Saprolegnia).

Formalin or Formalin based products can also be effective in bacterial infections, although primarily gram positive, which make up a small percentage of aquatic bacterial infections.

Dosage: 1 teaspoon of a 3% solution per ten gallons every other day for 10 days. Formalin combines well with malachite green and I recommend this combination for a more effective treatment.

Formalin can deplete oxygen in an aquarium (or pond) with a high bio load, so consider adding an air stone or a power head with an air diffuser. Formalin can be harsh on gill tissue, so be careful to not overdose.

As a bath 0.50 to 0.95 mLs per gallon (0.15 to 0.25 mLs per liter) for up to 60 minutes is an effective bath treatment (with less danger to the nitrifying bacterial bed).

Formalin is more toxic in soft, acidic water so buffering the water with a KH and GH of 50 ppm and 100 ppm respectively (or higher) is recommended. Regular Wonder Shells are an effective means to buffer a freshwater aquarium.

You have seen some anecdotal comments about formalin lowering dissolved oxygen levels, especially in ponds. This statement is based on a large die off minute life forms in the pond from treatment. This is incorrectly based on formalin being the only treatment that can cause this. Any treatment that kills protozoans or other similar life forms as well as many anti-bacterials can cause similar effects when the bio load is past what the pond can support. This is no more a problem with formalin than with any other treatment in a high bio load/ plankton environment.

Quick Cure is a very effective combination of Malachite Green and Formalin. A newer and in experience formulation that contains a formalin product is SeaChem ParaGuard which contains a proprietary, synergistic blend of aldehydes which is less toxic than other formalin/formaldehyde formulas. The form of Formalin used in ParaGuard is the safest/most effective formalin.

## Copper Sulfate

USE: For treatment of freshwater and marine ich (Crytocaryon), Oodinium, external parasites, fungus and even algae (especially in ponds). Copper Sulfate can be very effective when used properly and carefully.

Although not a first choice for freshwater ich (malachite green/formalin is generally more effective), Copper Sulfate is a very effective quarantine choice for ich and velvet prevention for new fish, especially when combined with Methylene Blue and used in a bare (no gravel) tank.

Copper used properly does not generally harm biological filters as quinine based treatments can (that are often used a copper substitutes in treating Marine Oodinium, Brooklynella, Cryptocaryon), however studies have shown bactericidal properties of Copper Sulfate on certain gram positive pathogens, so for this reason Copper should be used with care (or not at all in thanks that are not fully cycled.

Copper Sulfate products can be difficult to remove from marine aquariums that have crushed coral and coral based rock as coral skeletons and crushed coral readily absorb copper and then slowly leech it back out. For this reason copper based medications are often best used in hospital tanks when it comes to saltwater use. For freshwater use, this no where near as much a problem in my experience.

The more effective treatments for freshwater Velvet (Piscinoodinium pillulare), as copper will attack both swimming stage and the dormant stage (as this is when the parasite uses its chloroplasts to produce nutrients, and the copper destroys these).

Copper is the treatment of choice as far as effectiveness goes for Marine Ich (Crytocaryon) and Oodinium, unfortunately it cannot be used with invertebrates and often takes up to a month to remove traces from a marine display tank treated with copper before it is safe for invertebrates. For an alternative, Metronidazole has been shown to be an antibiotic that is generally effective for marine ich (although not as effective as copper and even less so against Oodinium).

Copper Sulfate such as Aquarisol is also effective for treatment of Shimmy in livebearers, especially mollies that are kept in low salt freshwater aquariums (the addition if positive mineral ions is also essential for the treatment of Livebearer Shimmy).

Dosage: Treat according to your solution to bring your copper level to .15 -.20 ppm for Oodinium/Brooklynella (and most freshwater applications) and .20-.25 ppm for Crytocaryon. Sequestered Copper (often called chelated, but that is incorrect, chelated means inactivated) sulfate works best (citric acid help achieve this). Soluble copper salts work well in freshwater only.

It is also noteworthy as to the use of copper in marine aquariums (especially tanks with any calcium based rock or substrate) that the amount of copper you will need to add will be high initially, but go down over subsequent days as no more copper is absorbed by the coral or even the minerals suspended in the water. This is important to note in treating display tanks, but is not as notable in bare hospital tanks; and for this reason copper should never be used in saltwater without a copper test kit on hand!

Please note that copper, basically kills the parasite by poisoning it more than the fish, so never overdose.

Do not use with snails and other invertebrates, do not use in reef aquariums, and note; when used as an algaecide, the copper is absorbed by the algae then released when it dies. Removal of sequestered copper can be difficult, only EDTA (Ethylene Diamine Tetra Acidic Acid) and water changes remove it, not carbon.

For an effective treatment for ich, fungus, and especially Velvet, Medicated Wonder Shells are extremely for Velvet & Livebearer Shimmy in particular as they have three of the most effective ingredients for the treatment of Velvet & Shimmy; Copper Sulfate, Acriflavin, & Methylene Blue. They also add electrolytes and calcium, essential for proper healing and osmoregulation.

## Potassium Permanganate KMNO4

The permanganate ion is a strong oxidizing agent. Potassium Permanganate is a strong oxidizer that when added to water will give a deep purple color that will slowly turn brown/yellow as it oxidizes (the more dissolved organics, the more quickly the change of color to brown/yellow which indicates the oxidation properties are mostly spent).

## Use

Being an oxidizer, Potassium Permanganate is useful for water clarification and odor elimination; often where a bacterial cloud is present Potassium Permanganate should only be used in established cycled aquariums). Potassium Permanganate can be used to reduce dangerous high organic levels in freshwater ponds and aquariums, especially "Bacterial Blooms" (Grey cloudy tanks).

It can be used as a special-purpose freshwater conditioner for ponds and aquariums, because of its ability to improve water quality by oxidation of excessive dissolved and suspended organics. The chronic presence of excessive dissolved organics in the water promotes the growth of potential disease-causing bacteria such as Aeromonas, and as well lower KH and pH. However it should be noted that initial oxidation produces carbon dioxide which will initially lower pH/KH and I recommend a water change a day after the use of potassium permanganate for the purpose of oxidation of organics in ponds or aquariums.

Probably, the best use for Potassium Permanganate is for fish baths or swabbing infections caused by bacterium such as Columnaris.

For these uses, a fish keeper would apply a direct swab to fish wounds or ulcers or as a bath at double recommended tank strength for half an hour.

Do not combine with Methylene Blue in baths. MB is a better choice where stress has occurred or ammonia and nitrite poisoning has occurred. Whereas potassium permanganate is the better choice for wounds/external, Flukes, external nematodes, Anchor worms, Columnaris or Aeromonas/Vibrio infections. However, even the use for Columnaris and the other before mentioned "diseases', it should be noted that is the fish are weakened severely or if the fish is very sensitive fish (such as scaleless fish like Knife fish), Methylene Blue may still be the better choice even if less effective, as the oxidizing properties of Potassium Permanganate may be much for these fish (half strength may be another option).

## Hydrogen Peroxide; H2O2:

Hydrogen Peroxide is a strong oxidizer with a Redox of +1.82.

Hydrogen peroxide is commonly used as a disinfectant for cleaning wounds in people, as well, Hydrogen peroxide has also been used in aquaculture to add oxygen to water (often prior to shipping) or as an immersion (bath) treatment against many different disease-causing organisms, including external parasites, bacteria, and fungi, on different species and life-stages of fish.

Since fish and other vertebrate animal cells produce one or another form of catalase (which is a Redox Reducer) for protection against effects of free radicals (which an oxidizer such as Hydrogen Peroxide is), dilute H2O2 is rapidly decomposed by catalase into oxygen and water, before it can do damage to the fish cell. That is the reason H2O2 will bubble when applied to an open cut.

However many lower animals such as shrimp do not appear to produce this catalase in enough quantity (if at all), so the use Hydrogen Peroxide in the presence of invertebrates such as Shrimp should be avoided.

Hydrogen Peroxide decomposes rapidly into water and oxygen, however in concentrations of Hydrogen Peroxide (not generally household), this can be explosive in a sealed container (due to concentrated oxygen buildup, not from hydrogen). This rapid decomposition into water and oxygen is what makes the use of Hydrogen Peroxide plausible considering its high oxidation properties. This is also how Hydrogen Peroxide is useful for the use in adding oxygen to fish prior to shipping (although care MUST be exercised so as to not over dose, which would be lethal.

As well continued use of Hydro Peroxide can stunt growth and as well result in more damage to a fish in the form of lower immunity than in help from oxidation of disease pathogens (due to high Redox Oxidation).

Household Hydrogen Peroxide is generally a solution of about three percent, while PEROX-AID® (Eka Chemicals, Marietta, Georgia) is 35%.

## USE/DOSAGE

Hydrogen Peroxide being a stronger oxidizer than chlorine can damage gills and other epidermal tissues on fish, especially adults. Gouramis and other Labyrinth fish seem to be especially sensitive to Hydrogen Peroxide and use with these fish should be avoided.

The use with eggs to prevent Saprolegnia is generally the safest, although baths for Saprolegnia and Columnaris treatment is another use, although not as safe based on my experience and research.

It is easy for the aquarist to involuntary introduce fish diseases to the aquarium . when purchasing new fishes, invertebrates, plants and aquarium decorations. Bringing new living organisms into the aquarium is naturally more risky than inserting a new piece of equipment, since most malevolent organisms rapidly succumb without a suitable host.

It is however possible for some organisms to stay dormant and survive without a host, and you should therefore always be careful when you move aquarium decoration and equipment from between aquariums.

White spot disease is caused by the Ichthyophthirius multifilis and the most distinguishing symptom is white spots that develop on the body and/or fins of your fish. Heavy breathing can be another symptom, and an infected fish will often scrub it self against rough surfaces in the aquarium.

Marine White Spot Disease is caused by a parasite named Cryptocaryon, but was earlier believed to be the result of an infection with a saltwater form of Ichthyophthirius. The symptoms are similar to freshwater Ich. It is possible for saltwater species become immune against Cryptocaryon and the immunity can stay for up to six months.

Skin and gill flukes are caused by parasitic flatworms known as Trematodes. Trematodes are called "flukes" since they have the same flattened body shape as a flounder, and the old Saxon name for flounder was fluke. A healthy fish will usually be able to withstand a small fluke attack, but a large fluke infestation can be lethal since skin flukes and gill flukes can cause severe tissue damage and lesions. Skin and gill flukes can also be the reason behind hyperplasia of both skin and gill epithelium. When the skin and gills are damaged by Trematodes, it will be even easier for new Trematodes to attack the fish.

Anchor worms are thread-like worms that attach themselves to the head of the fish. You need to treat your fish with a remedy especially made in order to kill Anchor worms, since common treatments such as marine salt will have little or no effect on the Anchor worms. Never pull out the worms using force, since the head of the worm will stay attached to the head of the fish and grow a new body.

Hole in the Head disease is caused by hexamita, a flagellated protozoan that is found in the gastrointestinal tracts of many fish species. Even healthy fish can carry Hexamita. If the water quality drops or if your fish is weakened by some other problem, Hexamita can spread from the intestines via feces in the water and attack the outside of the fish.

A small sore will form above the eyes of the fish and eventually grow in to a large hole. Large fish species are more prone to hole in the head disease; especially large cichlids such as Oscars and Jurupari Earth Eaters.

# FISH MEDICATIONS DOSING RECOMMENDATIONS

## Antibiotics

Amikacin sulfate (Amiglyde-V®). Loading dose of 5 mg/kg IM and then 2.5 mg/kg every 72 hours for 5 treatments total. (Source: Dr Greg Lewbart, 2008, Michigan Veterinary Conference Proceedings)

Ceftazidime (Fortaz®). Human cephalosporin drug with good broad-spectrum activity, long half-life, kills Pseudomonas. Dose is 20 mg/kg IM or SC Q 72 hours. (Source: Dr Greg Lewbart, 2008, Michigan Veterinary Conference Proceedings)

Enrofloxacin (Baytril®). May be used IM, ICe and PO at a dose of 5-10 mg/kg/Q 48 hours for 7-21 days. May be diluted with sterile saline since the concentrated form (2.27%) may be irritating. May also be used as a bath at a concentration of 2.5 mg/L for 5 hours Q 24 hours. Recent work (Lewbart et al., 2005) indicates 10 mg/kg ICe Q 5 days for koi at 20 degrees C. (Source: Dr Greg Lewbart, 2008, Michigan Veterinary Conference Proceedings)

Florfenicol. Based on salmonid literature, an estimated dose of 10-30 mg/kg administered IM every 24 hours would be required. This drug was recently (10/2005) approved for use in the U.S. in some food fish applications. (Source: Dr Greg Lewbart, 2008, Michigan Veterinary Conference Proceedings)

Trimethoprim sulfadiazine (Tribrissen®). A good broad-spectrum antibiotic. Dose is 30 mg/kg given IM, PO, or IC Q 24 hours for 7-10 days. May also be used as a bath at a concentration of 20 mg/L for 5 hours Q 24 hours. (Source: Dr Greg Lewbart, 2008, Michigan Veterinary Conference Proceedings)

Trimethoprim sulfamethoxazole (Bactrim, Septra, Aqua-sulfa) Potentiated sulfonamide is a bacterocidal antibiotic that inhibits protein synthesis. Broad spectrum antibiotic effective against both gram positive and gram negative bacteria (Most fish bacterial pathogens are gram negative), however, it may not be effective against pseudomonas. It is much preferable to plain sulfa antibiotics as the trimethroprim reduces bacterial resistance issues. The immersion dose is 20 mg/L dosed daily with a 50% water change prior to each dose. Dose for 5-7 days. (As with most antibiotics, dosing in medicated food at 30mg/kg food is preferred.

## Parasiticide

Dimilin (Diflubenzuron) Dose is 0.01 mg/L for 48 hours, 3 treatments, one week apart. Very effective but may kill desirable environmental invertebrates. Anecdotal reports exist where veterinarians are prescribing lufenuron (Program®) at a similar dose with good success. (Source: Dr Greg Lewbart, 2008, Michigan Veterinary Conference Proceedings)

Panacur (Fenbendazole) A good parasiticide for intestinal nematodes. Can be given orally at a dose of 50-100 mg/kg and repeated in two weeks. Also can be used in a gel food at a concentration of 0.2% (200 mg/100 grams food) and fed for three days, repeated in two weeks. (Source: Dr Greg Lewbart, 2008, Michigan Veterinary Conference Proceedings)

Formalin Bath dose is 20-25mg/L for 12 to 24 hours followed by a 50% water change. 1.0 ml of 100% formalin (37% formaldehyde)/ 10 gallons (38 L) of water. (Source: Dr Greg Lewbart, 2008, Michigan Veterinary Conference Proceedings)

Bath: 0.125-0.25 mL/L bath every 24hours for 2-3 day for up to 60 minutes, or .015-0.025 mL/L (15-25ppm) prolonged immersion, every two to three days. Change 50% water on nontreatment days. Is carcinogenic. (Source: Dr Helen Roberts, Veterinary Clinics: Exotic Animal Practice, Bacterial and Parasitic Diseases of Pet Fish)

Levamisole phosphate (Ripercol) 2 mg/L for 12-24 hours as a bath treatment. (Source: Dr Greg Lewbart, 2008, Michigan Veterinary Conference Proceedings)

Metronidazole (Flagy) Normally used as a bath treatment at 10 mg/L Q 24 hours for 3 consecutive days with 25-50% water changes between treatments. Also can be used in a gel food at a concentration of 0.2% (200 mg/100 grams food) and fed for ten days.

Bath: 7-15 mg/L every 24-48 hours for 5-10 days with a 50% water change between treatments. Orally dose is 25-50 mg/kg (0.25% in food fed at 1% body weight per day) for 3 days.

Praziquantel (Droncit) Dose is 5-8 mg/kg IM, IC, or PO repeated in 2 weeks. Bath treatment is 5-10 mg/L for 3-6 hours, repeat in 7 days.

Salt (sodium chloride) Safe for most species of ornamental fishes at a dose of one gram/L as an indefinite bath. Goldfish and koi will tolerate concentrations from 3-5 grams/L. (2 grams/L will usually kill circulating protozoal parasites. (Source: Dr Greg Lewbart, 2008, Michigan Veterinary Conference Proceedings)

## Sedatives

Clove Oil (Eugenol) Eugenol is not completely soluble in water and should be diluted 1:10 in 95% ethanol to yield a working stock solution of 100 mg/ml, since each ml of clove oil contains approximately 1 gram of drug. Concentrations of between 40 and 120 mg/L are effective in freshwater and marine species and results are comparable to MS-222, except that recovery may be prolonged.

Tricaine methanesulfonate (MS-222) Concentrations of between 100 and 150 mg/liter will usually anesthetize most fish within a matter of three to five minutes.

## Diuretics

Furosemide (Lasix) fish-0.5-2 mg/kg orally, IM, or IV (up to 8 mg/kg in emergencies.) (Source: University of Minnesota, Research Research Animal Resources, Veterinary Formulary).

## SALTWATER FISH DISEASES AND TREATMENTS

The most effective way to deal with diseases is to prevent it by minimizing stress.

According to Foster and Smith, "A varied diet, stable water temperature, good water parameters, daily observance, and responsible stocking are a few ways to ensure healthy fish, free of disease."

But even if you manage to deal with it, unfortunately it will still strike one or more of your saltwater fish. Fish will get sick.

Pathogens that cause diseases are in and around your fish's natural setting and in your aquarium. They may be bacterial, viral, fungal or parasitic.

So when the fish's skin or tissue gets damaged, whether from shipping, netting or any other circumstances, bacterial and fungal pathogens seize the opportunity to infect the fish.

Poor water quality also weaken the fish's immune system. This could be another cause for fish diseases.

The bad thing is that there's not a lot of treatment available for us aquarists and there is no guarantee that our saltwater fish will be saved.

We will cover topics that are most common in the home aquarium—how you'll be able to recognize these fish diseases and treatments available.

### *Bacterial Infections*

- Fin Rot
- Fish tuberculosis, wasting disease
- Vibriosis, ulcer disease

### *Viral Disease*

- Cauliflower or Lymphocystis

### *Fungal Disease*

- Ichthyophonus disease, whirling disease
- Exophiala disease

### *Parasitic Infestations*

- Marine velvet
- Marine white spot, cryptocaryoniasis, marine ich
- Uronema disease
- Tang turbellarian disease, black spot
- Tremode infestation
- Crustacean infestation

### *Other Health Problems*

- Head and lateral line errosion
- Poisoning

## Bacterial Infections

There are good and bad bacteria. And all bacteria are microscopic one-celled organisms are capable of reproduction at a very fast rate.

Remember that some bacteria are beneficial to the aquarium in the nitrogen cycle? But some can cause infections.

*Here is list of bacterial infection fish diseases and treatments.*

## Fin Rot

Causes: Aeromonas, Pseudomonas, Vibrio bacteria

Symptoms : This bacterial infection causes erosion or rotting of the fins and the fin rays. The base of the fins reddens. In an advance stage, it spreads to the skin causing bleeding to the gills.

Treatment : You can remove uneaten food, do a partial water change, and change the activated carbon in your filter.

The antibiotics furanace, augmentin, and ciprofloxin may be effective.

Recommended treatments: Melafix, Furanase, Furan-2, Maracyn, Maracyn-Two and Maracyn Plus.

## Fish Tuberculosis, Wasting Disease

Causes : Mycobacterium bacteria

Symptoms : There's really no external sign for this kind. If your fish looks fine from the outside appearance and looks healthy, it will be hard to recognize if it is internally infected.

Fish that are infected though may live a year or more. Skin lesions, emaciation, labored breathing, scale loss, frayed fins and loss of appetite are all signs of this infection. Unfortunately, by the time these symptoms are recognized, it is probably too late to save the fish.

Treatment : This bacteria are transmitted orally through raw infected fish flesh, detritus, and feces of the infected fish. They can also infect skin wounds and lesions.

The best treatment is prevention. Try not to feed raw fish and shellfish to your saltwater fish. Antibiotics, including kanamycin, erythromycin, and streptomycin can be used.

If the saltwater aquarium is heavily infected with this type, it must be sterilized and the water should be discarded.

## Vibriosis, Ulcer Disease

Causes : Vibriobacteria

Symptoms : Variety of symptoms are associated with this disease. They can include lethargy, darkening of color, anemia, ulcers on the skin and lower jaw, bleeding of the gills, skin and intestinal track, clouded eyes, loose scales, pale gills and sudden death.

Treatment : This bacteria commonly go straight to the intestinal tracts of a healthy fish. They become dangerous only when stress allows infection.

Poor water quality, crowding, excessive handling, and copper treatments are common causes of stress in saltwater aquarium fish. Immersion treatments with antibiotic compounds have met with some success.

## Viral Disease

Viruses are microscopic organisms that thrive by invading the cells of their hosts. In most cases, there are no treatments for the few viral diseases of the saltwater aquarium.

*Here is list of viral fish diseases and treatments.*

## Cauliflower or Lymphocystis

Causes : Cystivirusvirus

Symptoms : Fin and body lesions are raised, whitish, warty and have a lumpy texture like a cauliflower. It may take three to four weeks for these lesions to reach their full size.

The infection is not really fatal, but it can be transmitted to other fish in the saltwater aquarium.

Treatment : The only treatment is to isolate the fish immediately and let the fish's immune system deal with the infection. This may take as long as several months. Other hobbyist scrape the lesion off the fish.

## Fungal Disease

Fungi are plan-like organisms. Some are parasitic on fishes.

## Ichthyophonus Disease, Whirling Disease

Causes : Ichthyophonus fungus

Symptoms : These fungi goes to the internal organs, infecting the kidney, heart, spleen and liver.

Signs include emaciation, spinal curvature, darkening or paling of the skin, roughening of the skin, fin erosion, and skin ulcers. Erratic swimming behavior can be a symptom too.

Treatment : This parasitic organism has a complex life cycle. The fungal cysts are usually ingested by the fish, where thy burst and enter the bloodstream, infecting the internal organs.

Fishes that has this kind of disease typically die up to two months after being infected. Treatment is hard because of the internal nature of this disease.

The infected fish should be removed right away from the aquarium to prevent other fishes from becoming infected.

Recommended treatments: Rid-Ich+ or Probiotic Marine Formula. For scaless fish: Rid-Ich+ or Ich-Attack.

## Exophiala Disease

Causes : Exophialafungus

Symptoms : Lethargy, disorientation, and abnormal swimming are the signs.

Treatment : No treatment is known. The best thing to do is to isolate the fish to prevent others fishes from contracting the fungus.

## Parasitic Infestations

### *Marine velvet*

Causes : The dinoflagellate protozoan Amyloodinium ocellatum

Symptoms : You can usually see the infection on the gills then it spreads to the skin. It becomes dull, patchy and velvet-like. White spots are visible on section of intact skin.

The fish's behavior may include fasting, gasping, scratching against objects, and sluggishness. Death can occur in as little as 12 hours if infected. Tangs, clownfish, and angelfish are especially susceptible.

Treatment : The organism has three stages to its life cycle and only one of them is parasitic. No effective treatment is known, but copper-based products and some antibiotics are effective.

A freshwater dip sometimes digs out the parasites from the host but does not kill them. Treatments are often dragged and the entire aquarium must be treated to fully destroy the infestation.

### Marine white spot, cryptocaryoniasis, marine ich

Causes : The ciliate protozoan Cryptocaryon irritans

Symptoms : Early signs are fasting, cloudy eyes, troubled breathing, excess skin mucus and pale skin. White spots appear on the skin, gills, eyes and death usually follows within a few days.

Treatment : White spot organisms are hard to control. The freshwater dip may be effective in killing parasite on the fish but does little to the aquarium. Prolonged immersion in copper products is the most effective treatment.

### Uronema disease

Causes : The ciliate protozoan Uronema marinum

Symptoms : External ulcers, muscle and skin bleeding, lethargic behavior, cloudy eyes, and sloughing of the skin are signs. They usually die right away if infected because of the impaired circulation of the gills.

Treatment : Freshwater bath followed by prolonged immersion in a formalin bath a day later is effective.

### Tang turbellarian disease, black spot

Causes : Paravortex flatworms

Symptoms : Even if the name implies that only tangs are infected, this is not the case. Many can be infected by this flatworm.

The early phase, the organism would look like numerous dark spots over the fins, gills, and body. Other signs include fastening, listlessness, paling or whitish skin, and scratching against objects.

Treatment : As most parasitic infestations, crowding allows the disease to spread to other fish in the aquarium. Freshwater bath, triclorfon, formalin or praziquantel immersion can be effective.

## Tremode infestation

Causes : Monogenetic trematode worms

Symptoms : Many species of this worms are too small to see without a microscope. They normally infect gills, eyes, skin, mouth and anal opening.

The infected fishes usually rub themselves against objects in the aquarium, trying to remove these parasites. These often causes damage that leads to bacterial infections.

Treatment : Immersion in fresh water, mebendazole, praziquantel or trichlorion is effective against trematodes.

## Crustacean infestation

Causes : Copepod, isopod, and argulid crustaceans

Symptoms : Most of these tiny, crab-like organisms are visible to the naked eyes. Copepods remain fixed in the same position while isopods are argulids move over the surface of the fish, causing tissue damage.

Fishes that are infected swim erratically, rub against objects in the aquarium and jump. Bacteria infect lesions.

Treatment : Remove the ones that are infected immediately. Remove the decorations as well and dry them to kill the egg masses or immerse them in 2 percent bleach solution for two hours.

Treat the infected fish with fresh water, trichlorfon or malathion baths.

## Other Health Problems

Just when you thought there could be no more, here are a couple more that you need to be aware of.

## Head and Lateral Line Errosion

Causes : Poor water quality, nutrient deficiency, possibly a parasite

Symptoms : Like the freshwater disease called hole-in-the-head, holes develop and enlarge in the sensory pits of the head and down the lateral line on the body.

The disease progress slowly and the fish does not seem to behave differently. Advanced stages can lead to bacterial infection and death.

Treatment : There are no specific treatment, but some recommended the use of the freshwater antibiotic flagyl. Check your water quality and make the necessary adjustments. Make sure that you are also meeting the fish's nutrional needs. Get a variety of food for their diet and add vitamin supplements to their food.

## Poisoning

Causes : Multiple causes including buildup of nitrogenous compounds (ammonia, nitrite), household chemicals (smoke, cleaners, fumes) and tap water constituents (heavy metals, chloramines)

Symptoms : Low levels of toxins in the water stress the fish that can lower their resistance.

Higher levels causes abnormal behavior such as darting movements, jumping, and gasping at the surface of your aquarium water.

Treatment : Be sure that activated carbon is used to remove toxins. Do a water change, about 20-40% of water should be replaced.

# BIBLIOGRAPHY


- Allerston CK, Shimizu M, Fujieda M, Shephard EA, Yamazaki H, Phillips IR (2007). "Molecular evolution and balancing selection in the flavin-containing monooxygenase 3 gene (FMO3)". Pharmacogenet Genomics. 17 (10): 827-39. doi:10.1097/FPC.0b013e328256b198. PMID 17885620.
- Anderson B.G.: Atlas of Trout Histology, Wyoming Department of Fish and Game, 1974.
- Durborow, R.M., and R. Francis-Floyd. 1996. Medicated feed for food fish. Southern Regional Aquaculture Center Publication 473, Mississippi State Univ. MS.
- Federal Joint Subcommittee on Aquaculture. 1994. Guide to Drug, Vaccine, and Pesticide Use in Aquaculture. Texas Agriculture Extension Service, College Station, TX.
- Ferguson H.W.: Systemic Pathology of Fish, Iowa State Press, Ames, Iowa, 1989.
- Furnes B, Feng J, Sommer SS, Schlenk D (2003). "Identification of novel variants of the flavin-containing monooxygenase gene family in African Americans". Drug Metab Dispos 31 (2): 187-93. doi:10.1124/dmd.31.2.187. PMID 12527699.
- Hernandez D, Addou S, Lee D, Orengo C, Shephard EA, Phillips IR (2003). "Trimethylaminuria and a human FMO3 mutation database". Hum Mutat 22 (3): 209-13. doi:10.1002/humu.10252. PMID 12938085.
- Humbert JA, Hammond KB, Hathaway WE. (1970). "Trimethylaminuria: the fish-odour syndrome". Lancet 2 (7676): 770-1. doi:10.1016/S0140-6736(70)90241-2. PMID 4195988.
- Mitchell SC, Smith RL (2001). "Trimethylaminuria: the fish malodor syndrome". Drug Metab Dispos 29 (4 Pt 2): 517-21. PMID 11259343.
- Roberts R.J: Fish Pathology, Bailliere Tindall, London, Second edition, 1989.
- Rottman, R. W., R. Francis-Floyd, and R. Durborow. 1992. The role of stress in fish disease. Southern Regional Aquaculture Center Publication 474, Mississippi State Univ. MS.
- Rottman, R. W., R. Francis-Floyd, P.A. Reed, and R. Durborow. 1992. Submitting a sample for fish kill investigation. Southern Regional Aquaculture Center Publication 472, Mississippi State Univ. MS.
- Treacy E, Johnson D. Pitt JJ and Danks DM (1995). "Trimethylaminuria, fish odour syndrome: A new method of detection and response to treatment with metronidazole". J Inherit Metab Dis 18 (3): 306-12. PMID 7474897.

- Treacy EP, et al. (1998). "Mutations of the flavin-containing monooxygenase gene (FMO3) cause trimethylaminuria, a defect in detoxication". Human Molecular Genetics 7 (5): 839-45. doi:10.1093/hmg/7.5.839. PMID 9536088.
- Yamazaki H, Fujieda M, Togashi M, et al. (2004). "Effects of the dietary supplements, activated charcoal and copper chlorophyllin, on urinary excretion of trimethylamine in Japanese trimethylaminuria patients". Life Sci. 74 (22): 2739-47. doi:10.1016/j.lfs.2003.10.022. PMID 15043988.
- Zschocke J, Kohlmueller D, Quak E, Meissner T, Hoffmann GF, Mayatepek E (1999). "Mild trimethylaminuria caused by common variants in FMO3 gene". Lancet 354 (9181): 834-5. doi:10.1016/S0140-6736(99)80019-1. PMID 10485731.

# INDEX